Olfa Kanoun (Ed.)
Progress Reports on Impedance Spectroscopy

Also of Interest

Low Power VLSI Design
A. Sarkar, S. Dep, M-Chanda, S.K. Sarkar, 2016
ISBN 978-3-11-045526-7, e-ISBN 978-3-11-045529-8,
e-ISBN (EPUB) 978-3-11-045545-8, Set-ISBN 978-3-11-045555-7

Chaotic Secure Communication
K. Sun, 2016
ISBN 978-3-11-042688-5, e-ISBN 978-3-11-043406-4,
e-ISBN (EPUB) 978-3-11-043326-5, Set-ISBN 978-3-11-043407-1

Nano Devices and Sensors
J. J. Liou, S.-K. Liaw, Y.-H. Chung (Eds.), 2016
ISBN 978-1-5015-1050-2, e-ISBN 978-1-5015-0153-1,
e-ISBN (EPUB) 978-1-5015-0155-5, Set-ISBN 978-1-5015-0154-8

Systems, Automation and Control
N. Derbel (Ed.), 2016
ISBN 978-3-11-044376-9, e-ISBN 978-3-11-044843-6,
e-ISBN (EPUB) 978-3-11-044627-2, Set-ISBN 978-3-11-044844-3

Power Electrical Systems
N. Derbel (Ed.), 2016
ISBN 978-3-11-044615-9, e-ISBN 978-3-11-044841-2,
e-ISBN (EPUB) 978-3-11-044628-9, Set-ISBN 978-3-11-044842-9

Progress Reports on Impedance Spectroscopy

―

Measurements, Modeling, and Application

Edited by
Olfa Kanoun

Editor
Prof. Olfa Kanoun
Technische Universität Chemnitz
Faculty of Electrical Engineering and Information Technology
Reichenhainer Str. 70
09126 Chemnitz, Germany
Olfa.Kanoun@etit.tu-chemnitz.de

ISBN 978-3-11-044756-9
e-ISBN (PDF) 978-3-11-044982-2
e-ISBN (EPUB) 978-3-11-044767-5
Set-ISBN 978-3-11-044983-9

Library of Congress Cataloging-in-Publication Data
A CIP catalog record for this book has been applied for at the Library of Congress.

Bibliographic information published by the Deutsche Nationalbibliothek
The Deutsche Nationalbibliothek lists this publication in the Deutsche Nationalbibliografie;
detailed bibliographic data are available on the Internet at http://dnb.dnb.de.

© 2017 Walter de Gruyter GmbH, Berlin/Boston
Typesetting: Konvertus, Haarlem
Printing and binding: CPI books GmbH, Leck
♾ Printed on acid-free paper
Printed in Germany

www.degruyter.com

Preface

Impedance spectroscopy is a widely used and powerful measurement method applied in many fields of science and technology such as electrochemistry, material science, biology and medicine. In spite of the apparently different scientific and application background in these fields, applications share the same measurement method in a system identification approach and profit from the possibility to use complex impedance over a wide frequency range and giving interesting opportunities for separating effects, for accurate measurements and for simultaneous measurements of different and even non-accessible quantities.

For electrochemical impedance spectroscopy (EIS) competency from several fields of science and technology is indispensable. Understanding electrochemical and physical phenomena is necessary for developing suitable models. Suitable measurement procedures should be developed taking specific requirements of the considered application into account. Signal processing methods are very important for extracting target information by suitable mathematical methods and algorithms. New trends are emerging rapidly involving special techniques for realizing fully automatic embedded solutions at low costs and requiring a deep overview of modern information technology.

The scientific dialogue between specialists of impedance spectroscopy, dealing with different fields of science, technology and application, is therefore particularly important to promote the adequate use of this powerful measurement method in both laboratory and in embedded solutions.

Since 2008, the International Workshop on Impedance Spectroscopy (IWIS) has been launched as a platform for promoting experience exchange and networking in the scientific and industrial field. Its aim is to serve for encouraging the sharing of experiences between scientists and to support new comers aiming to specialize in impedance spectroscopy. Since many years, the workshop has been gaining increasingly more acceptance in both scientific and industrial fields and addressing increasingly more fundamentals, but also diverse application fields of impedance spectroscopy. Many renowned scientists are contributing yearly to it and sharing their experience with scientists all around the world. By means of tutorials and special sessions, young scientists get a good overview of different fundamental sciences and technologies helping them to get expertise even in fields, which are still not in the focus of their background.

In 2013 the Circle of Experts of Impedance Spectroscopy (CEIS) was founded to promote exchange between experts and together with industry as interest group for promoting impedance spectroscopy all over the subfields related to fundamental and applications of impedance spectroscopy. The CEIS is the steering committee of the IWIS workshop supporting it and deciding about the yearly best paper and best poster award to recognize best contributions.

This peer reviewed book is the first edition in the series Progress Reports on Impedance Spectroscopy which has the aims to widen knowledge of scientists in this field by presenting selected and extended contributions from the International Workshop on Impedance Spectroscopy (IWIS'14 and IWIS'15). The series reports about new advances and different approaches in dealing with impedance spectroscopy, including theory, methods and applications. The book is therefore interesting for researcher and developers in the field of impedance spectroscopy.

I thank all contributors for the interesting contributions and the reviewers who supported by the decision about publication with their valuable comments.

Prof. Dr.-Ing. Olfa Kanoun

Contents

Part I: Batteries

Florina Cuibus, Svetlozar Ivanov, Ulf Schwalbe, Marco Schilling and Andreas Bund
State-of-Charge and State-of-Health Estimation of Commercial LiFePO$_4$ Batteries by means of Impedance Spectroscopy —— 3
1	Introduction —— 3	
2	Experimental Set-up —— 5	
3	Results and Discussion —— 6	
3.1	SoC Measurements —— 6	
3.2	SoH Measurements —— 9	
3.3	Data Modelling —— 11	
4	Conclusions and Outlook —— 16	
5	References —— 17	

Delf Kober, Oliver Görke and Julia Kowal
Ageing Investigation of Lithium Ion LiFePO$_4$ Batteries with a Combination of EIS and Structural Analysis —— 19
1	Introduction —— 19	
2	Experimental —— 20	
3	Results —— 21	
3.1	Electrochemical Impedance Spectroscopy —— 21	
3.2	Post-mortem Analysis —— 23	
4	Summary and outlook —— 27	
5	References —— 27	

Paul Büschel, Thomas Günther and Olfa Kanoun
Streamlining Calculation of the Distribution of Relaxation Times from Time Domain Data —— 29
1	Introduction —— 29	
2	Novel Algorithm —— 30	
3	Experimental —— 32	
4	Conclusion and Outlook —— 34	
5	References —— 34	

Christian Reinke, Kristian Nikolowski, Mareike Wolter and Alexander Michaelis
Influence of the Anode Graphite Particle Size on the SEI Film Formation in Lithium-Ion Cells —— 35
1 Introduction —— 35
2 Experimental —— 36
3 Results —— 39
4 Conclusion —— 41
5 References —— 42

Thomas Günther and Olfa Kanoun
Frequency-Dependent Phase Correction for Impedance Measurements —— 44
1 Introduction —— 44
2 Method —— 45
2.1 Measurement Model —— 45
2.2 Phase Correction —— 46
3 Experimental Validation —— 47
4 Conclusion —— 48
5 References —— 48

Sebastian Socher, Claudius Jehle and Ulrich Potthoff
On-line State Estimation of Automotive Batteries using In-situ Impedance Spectroscopy —— 49
1 Introduction —— 49
2 Experimental Details —— 50
3 Results and Discussion —— 50
3.1 Impedance Measurements —— 50
3.2 In-situ Impedance Measurements —— 52
4 Conclusions —— 54
5 References —— 54

Part II: Sensors

Christian Weber, Markus Tahedl and Olfa Kanoun
Capacitive Measurements for Characterizing Thin Layers of Aqueous Solutions —— 59
1 Introduction —— 59
2 Measurement Set-up —— 60
3 Data Validation —— 62
4 Experimental Investigations —— 64
5 Results —— 67
6 Conclusion —— 70
7 References —— 71

Łukasz Macioszek and Ryszard Rybski
Low-Frequency Dielectric Spectroscopy Approach to Water Content in Winter Premium Diesel Fuel Assessment —— 73

1	Introduction —— 73	
2	Material and Methods —— 74	
2.1	Diesel Fuel Samples —— 75	
2.2	Electrodes —— 75	
2.3	Impedance Spectroscopy —— 76	
3	Results and Discussion —— 76	
4	Conclusion —— 79	
5	References —— 80	

Christian Weber, Markus Tahedl and Olfa Kanoun
A Novel Method for Capacitive Determination of the Overall Resistance of an Aqueous Solution —— 81

1	Introduction —— 81	
2	Proposed Measurement Method —— 82	
3	Impedance Simulation —— 83	
4	Experimental Verification —— 86	
5	Conclusion —— 88	
6	References —— 88	

Part III: Material Characterization

Bernhard Roling and André Schirmeisen
Nanoscale Electrochemical Characterization of Materials by means of Electrostatic Force and Current Measurements —— 91

1	Introduction —— 91	
2	Experimental —— 94	
3	Results and Discussion —— 95	
3.1	Time-Domain Electrostatic Force Spectroscopy (TD-EFS) —— 95	
3.2	Nanoscale Impedance Spectroscopy —— 99	
4	Conclusion —— 102	
5	References —— 103	

Abdulkadir Sanli, Abderrahmane Benchirouf, Christian Müller and Olfa Kanoun
AC Impedance Investigation of weg Multi-walled Carbon Nanotubes/PEDOT:PSS Nanocomposites Fabricated with Different Sonication Times —— 105

1	Introduction —— 105	
2	Experimental —— 106	
2.1	Materials and Nanocomposite Preparation —— 106	
2.2	Measurement Set-up —— 108	

3	Results and Discussions —— 108
3.1	Morphological Analysis —— 108
3.2	Electrical Characteristics —— 109
3.3	Equivalent Circuit Modelling —— 110
4	Conclusions and Outlook —— 114
5	References —— 114

Part IV: Bioimpedance

Marco Carminati
From Counting Single Biological Cells to Recovering Photons: The Versatility of Contactless Impedance Sensing —— 119

1	Introduction —— 119
2	A General Equivalent Impedance Model —— 120
3	Impedance Sensing in Cell Biology —— 122
4	Impedance-Based Light Monitoring —— 123
5	Common Design Criteria —— 126
6	Conclusions —— 128
7	References —— 129

Hip Kõiv, Ksenija Pesti and Rauno Gordon
Electric Impedance Measurement of Tissue Phantom Materials for Development of Medical Diagnostic Systems —— 131

1	Introduction —— 131
2	Materials and Methods —— 132
2.1	Gelatine Phantom Preparation —— 132
2.2	Plexiglas Enclosure with Electrodes —— 133
2.3	Van der Pauw Method —— 133
2.4	Measurement of Gelatine Sample Conductivity —— 135
3	Results —— 135
4	Conclusion —— 136
5	References —— 137

Paco Bogónez-Franco, Pascale Pham, Claudine Gehin, Bertrand Massot, Georges Delhomme, Eric McAdams and Regis Guillemaud
Problems Encountered during Inappropriate Use of Commercial Bioimpedance Devices in Novel Applications —— 138

1	Introduction —— 138
2	Materials —— 140
2.1	Commercial Impedance Analyzers —— 140

2.2	Electrode–Skin Contact Impedance Measurement —— **140**	
2.3	Tissue Impedance measurement —— **141**	
2.4	Electrode–Skin Contact Impedance Electrical Model —— **141**	
2.5	Tissue Electrical Model —— **142**	
2.6	Fitting Impedance Measurements —— **142**	
2.7	Errors —— **142**	
2.8	Measurements on Healthy People —— **143**	
3	Experimental Results —— **143**	
3.1	Effect of Contact Impedance —— **143**	
3.2	Effect of Contact Impedance Mismatch —— **144**	
4	Discussion —— **150**	
5	Conclusion —— **151**	
6	References —— **152**	

Part I: **Batteries**

Florina Cuibus, Svetlozar Ivanov, Ulf Schwalbe, Marco Schilling and Andreas Bund

State-of-Charge and State-of-Health Estimation of Commercial LiFePO$_4$ Batteries by means of Impedance Spectroscopy

Abstract: Commercial LiFePO$_4$-type cells were characterized by electrochemical impedance spectroscopy to identify sensitive and reliable parameters suitable to diagnose the state of charge and state of health. On the basis of the proposed equivalent circuit, electrochemical parameters were extracted, and physical and chemical ageing phenomena were categorized and selected. It was observed that, after 2,000 cycles the charge transfer resistance increased approximately with 50% from the initial value, which indicates a decreasing of the exchange current density. The Warburg element showed a decrease of diffusion coefficient and lithium concentration due to the formation of solid electrolyte interface layer. The charge transfer resistance demonstrates more significant trend with ageing cycles and the results are consistent with experiment-based observations from the literature, which seems to be indicating the potential of the proposed model for battery age estimation.

Keywords: LiFePO$_4$ batteries, state of charge, state of health, electrochemical impedance spectroscopy

1 Introduction

The improvement of the storage system is one of the primary challenges for the development of modern electric vehicles. To fulfil the requirements for extended driving range, the storage systems need reliable high-energy batteries and an advanced battery management system (BMS) for efficient energy consumption optimization. The implementation of an efficient battery state-of-charge (SoC) and state-of-health (SoH) estimation is a central aspect for the development of BMS.

Battery ageing responsible for the cell impedance increase and power decay and energy decay origins from multiple and complex mechanisms [1]. Material properties, as well as storage and cycling conditions, have an impact on battery lifetime performance SoC and SoH. Therefore, prompt and reliable SoC and SoH determination, independent on the cycling current and operational history, is highly favourable. The

---

Florina Cuibus, Svetlozar Ivanov and Andreas Bund, Electrochemistry and Electroplating Group, Ilmenau University of Technology, 98693 Ilmenau, Germany
Ulf Schwalbe and Marco Schilling, Industrial Electronics Group, Ilmenau University of Technology, 98693 Ilmenau, Germany

DOI 10.1515/9783110449822-001

SoC, battery instability and ageing phenomena can be investigated non-destructively by means of electrochemical impedance spectroscopy (EIS) [2–4]. EIS is an established method for analysis of electrochemical systems and is considered a valuable tool to detect changes in mass transport properties, double-layer capacitance, ohmic resistance and reaction kinetics during battery operation and ageing. Furthermore, to the extensive use of this method add the properties of the cell (capacity, voltage), although the advantages of EIS method, the determination of electrical parameters of the individual electrode materials in commercial batteries, remain challenging. The implementation of EIS method (quick and online) or the determination of SoC and SoH for the BMS becomes as well challenging. The advantages of the EIS technique are that, the method does not have a pre-history and is sensitive enough for electrical parameter extraction. Therefore, an efficient estimation of SoC and SoH will improve the performance of the battery, developing better BMS. However, the SoC and SoH estimation is generally limited, as for example by other conventional methods like coulomb counting, which is not able to give the resistance, a crucial parameter for ageing diagnosis. In order to have an online and broad diagnosis, the resistance of material and electrolyte can be determined via EIS by an appropriate model extracted. Furthermore, EIS remains the most promising non-destructive method even if less practical application for battery characterization was employed. The capability of EIS measurements was already studied by Mingant et al. [5] by developing a diagnosis tool for SoC and SoH characterization. A lifetime model proposed by Omar et al. [6] provides a clear view regarding the change of the internal resistance and capacity fade at different conditions. On the basis of the parameters' evolution, a cycle life model was developed, which was proposed for the development of an accurate estimation of SoH. A series of EIS measurements performed in galvanostatic mode showed that the SoC and SoH have an influence on EIS. Good reviews of the available methods for SoC estimation are presented in [7, 8]. A large number of different types of Li-ion batteries (LiBs) existing on the market have advantages and disadvantages related to cost, performance, safety and cycle life. Considering safety and production cost, one of the suitable cell types is the $LiFePO_4$/graphite cell [9, 10]. The $LiFePO_4$ (LFP) cathode material received attention because of its long cycle and calendar life, low price and good thermal stability. Further advantages like fast kinetics (high power), low thermal heat exchange, stable structure, low price and non-toxicity, used for electrical cars, are the reasons why the LFP battery was chosen for this study.

The primary aim of this study is to characterize commercial LFP-type cells using EIS. Key aspects are SoH and SoC analysis by interpretation of the EIS spectral data, targeting identification of sensitive and reliable parameters, suitable for expressing SoC and SoH battery diagnostics. To perform specific ageing tests, physical and chemical ageing phenomena were categorized and selected. Further aim is to find appropriate scalable electrical parameters for simultaneous evaluation of SoC and SoH.

2 Experimental Set-up

Commercial 18650 cells (A123 Systems cylindrical) with a nominal capacity of 1.1 Ah and an LFP-based positive electrode were employed for all the measurements. The nominal voltage is 3.3 V and voltage limits are 3.6 and 2 V. The measurements were carried out using a BioLogic Potentiostat/Galvanostat (type VSP, France). To test the influence of various factors on the ageing of a commercial LFP cell, a test procedure shown in Fig. 1 was chosen. Concomitantly the SoC and SoH parameters based on EIS measurements were extracted and further modelled to predict the ageing of the battery.

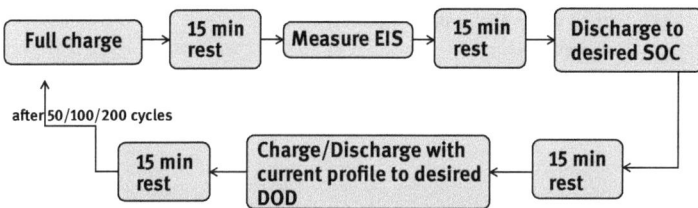

Fig. 1. Test procedure employed for SoC and SoH determination by means of EIS.

The first row describes the experimental procedure for SoC data acquisition by means of EIS. The procedure consists of EIS measurements, resting time, charge–discharge cycling current and discharge profile. The resting time of 15 minutes was experimentally determined. Experiments have shown that a resting time of 15 minutes is sufficient to have minimal influence on the capacity and parameter determination in the SoC range of interest. A charging current of 1.5 A and a discharging current of 1 A were chosen and represent the recommended charging–discharging current profile from the data sheet of the cells [11]. The cells were charged and discharged with 100% depth of discharge (DoD), which represents 3.6 V, the end-of-charge voltage, and 2 V, the cut-off voltage. EIS was measured at different SoC and SoH at OCP (I = 0 A) and a current amplitude of 50 mA. The EIS spectra were recorded between the frequency range 100 kHz and 10 mHz. The EIS test procedure is in Fig. 2 described.

EIS was performed for each cell at successive discharge intervals until a 2-V cut-off voltage was reached. The SoC of each cell was calculated on the basis of nominal voltage of the cell. In this study, the galvanostatic charge–discharge cycling of cells was conducted using a BioLogic Potentiostat/Galvanostat (type VSP, France) at room temperature (about 25°). The settings of voltage range and current density were chosen in correlation with the desired DoD.

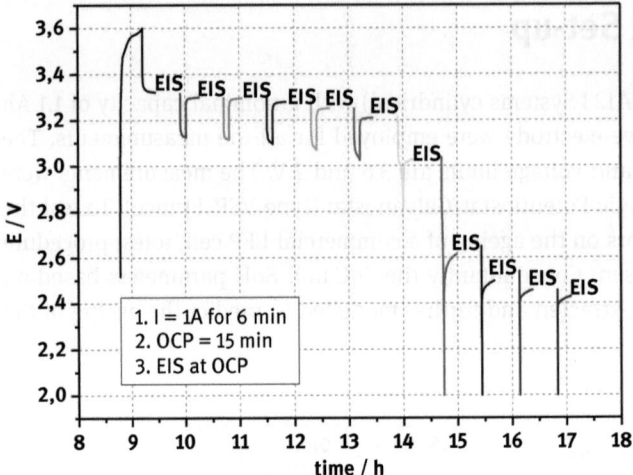

Fig. 2. EIS test procedure for different SoCs.

3 Results and Discussion

The goal of the electrochemical impedance tests is to verify the capability of EIS measurements and to propose an electrical circuit based model which will reveal the ageing effects of an LFP commercial cell.

3.1 SoC Measurements

EIS measurements were performed on A123 cylindrical (APR18650) LFP batteries at SoC ranging from 90% to 10% within the frequency range 10 mHz to 100 kHz. The EIS spectra were recorded after each 10% SoC, and the changes of these parameters for each battery type were observed. A typical Nyquist plot of the experimental impedance data is presented in Fig. 3.

As can be seen in Fig. 3 impedance curves in the region of less than 30% SoC are apparently different from others. At small SoC the ageing is not following a linear degradation, and this is seen by the recorded EIS. During practical use the battery does not perform in a low SoC range and therefore the extreme regions of SoC will not be taken in consideration for modelling. Moreover, the modelling of the extreme regions will increase the model complexity and calculation time; thus it is better to find model parameters in the middle of the SoC region and to include the model error in the extreme SoC region. Regarding Bode plot, very small impedance changes (≤1 mΩ) were observed (Fig. 4). It can be seen that an accurate model needs to be developed to predict the changes of the parameters in a precisely manner. The model should be accurate enough to record the very small impedance changes that are taking place at different SoCs in the battery.

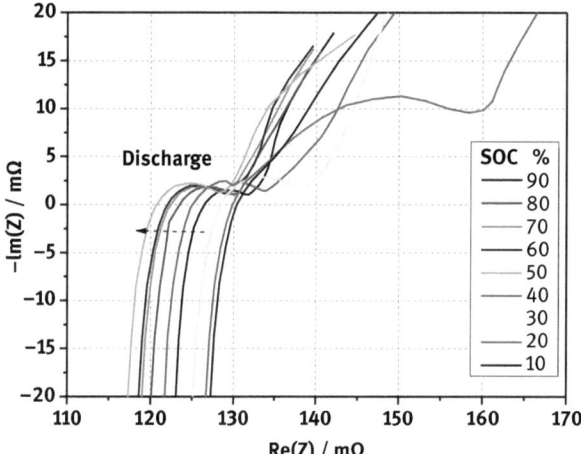

Fig. 3. EIS spectra of the APR18650 LFP 1.1 Ah battery at different SoCs.

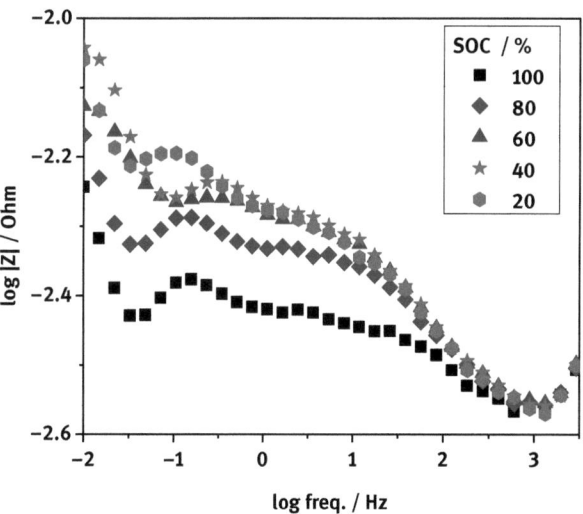

Fig. 4. Typical Bode plot of a 1.1-Ah LFP battery at different SoCs.

Interesting frequency range for SoC estimation represents 0 Hz to 60 kHz, and will be further used for SoC implementation in the BMS. Furthermore, for different SoC levels, behaviour should be checked if one developed model will fit the data from batteries of other geometries and configurations. In accordance to our idea it was confirmed that, different EIS spectra could be modelled with the same equivalent circuit and slight differences in the electrode resistance components were observed [12].

A correlation between the cells' imaginary impedance component and the SoC at three different frequencies was performed. The imaginary component of the cell impedance at 32.5 mHz, 1.1 Hz and 125.2 Hz versus SoC is displayed in Fig. 5. At all the three frequencies and the lowest SoC (20% and 10%) the imaginary component of the impedance varies independently with SoC. For the higher SoC (30%–90%) the imaginary impedance component remains constant, showing a linear behaviour. For a better explanation of our results, the literature presents different approaches. One approach is fuzzy logic modelling, which shows that, over the investigated SoC range the impedance data vary linearly describing the possibility of further mathematical analysis. The prediction of SoC and SoH for different battery systems with implementation of the prototype into the commercial hardware was aimed [13]. Another approach which describes the results and was developed to avoid ambiguous results was presented by Schänleber et al. [14]. By using the Kramers–Kronig test form, it was shown that by choosing wrong number of RC elements the simulation can fail delivering wrong results. A strategy was proposed to serve to a quick test, revealing the defectiveness of the measured impedance spectra. The correlations of the above-presented approaches with our results show that the behaviour of the impedance data parameters can be further investigated by mathematical analysis.

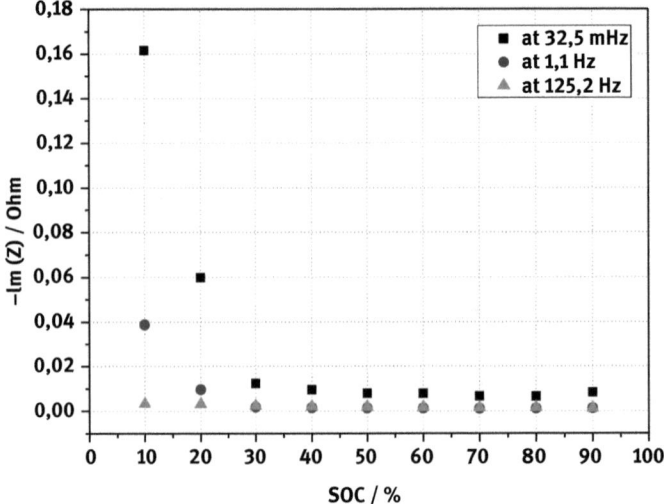

Fig. 5. Imaginary impedance component at different frequencies versus SoC for 1.1 Ah LFP battery.

3.2 SoH Measurements

The best indicator for decreasing the state of health (SoH) is the capacity monitoring. A capacity decrease with about 18% from initial capacity value over the 1,000 cycles was observed for the 1.1 Ah battery (Fig. 6). The decrease of capacity with cycle number was attributed to different ageing effects that are taking place in the battery. Furthermore, the capacity fade mechanism is predominantly attributed to the loss of lithium quantity and active material. These results were confirmed by the works of Dubarry et al. [15, 16]. Different measurements were conducted to quantify capacity loss and peak power capability degradation with cycle number to the end-of-life of the battery. They observed that the main cause of capacity fade is the loss of lithium inventory and active material. Moreover, when the capacity fade is controlled by the loss of active materials, the anode needs to be charged to a higher SoC and this outpaces the loss of lithium inventory in the system. Thus, the remaining active carbon material in the anode needs to be charged to a higher SoC to accommodate the lithium ions transferred from the cathode [17].

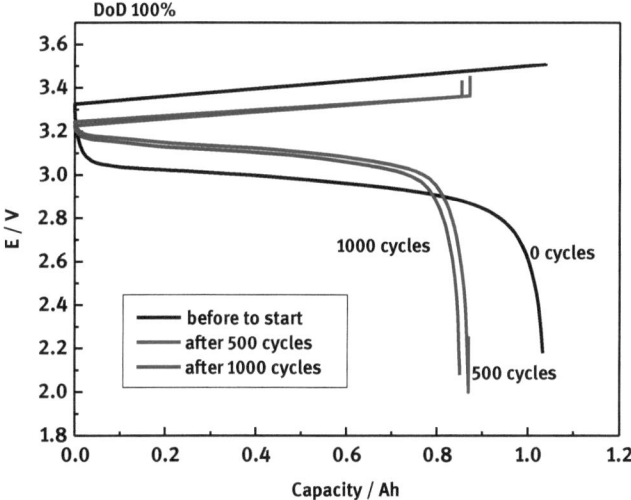

Fig. 6. Evolution of voltage versus capacity during cycling for 1.1 Ah LFP battery.

EIS measurements performed for 1.1 Ah LFP battery showed evolution of the impedance frequency behaviour upon change of SoH (Fig. 7A). For a better overview on the impedance changes versus frequency, typical Bode diagrams at different SoHs are presented in Fig. 7B.

Significant variations of imaginary impedance component, total impedance ($|Z|$), were observed over the entire frequency range. The impedance of the cell increases with increasing the cycle number, while the phase remains constant with changing the cycle number. It was concluded that the ageing influences the impedance spectra in the frequency range between 0.01 and 100 Hz. During cycling of an LFP battery, a volume change of approximately 6.77% was reported [18, 19]. With the insertion and extraction of lithium ions the cathode material gradually transforms from $FePO_4$ into $LiFePO_4$ corresponding to a phase transformation region of the cathode. This phase transition change brings cracks and loss of lithium, which at the end are as ageing effects counted. To clarify the ageing effects, the impedance data have been interpreted by a simplified equivalent circuit model, taking into account the effects of the charge transfer, double-layer capacitance and transport phenomena.

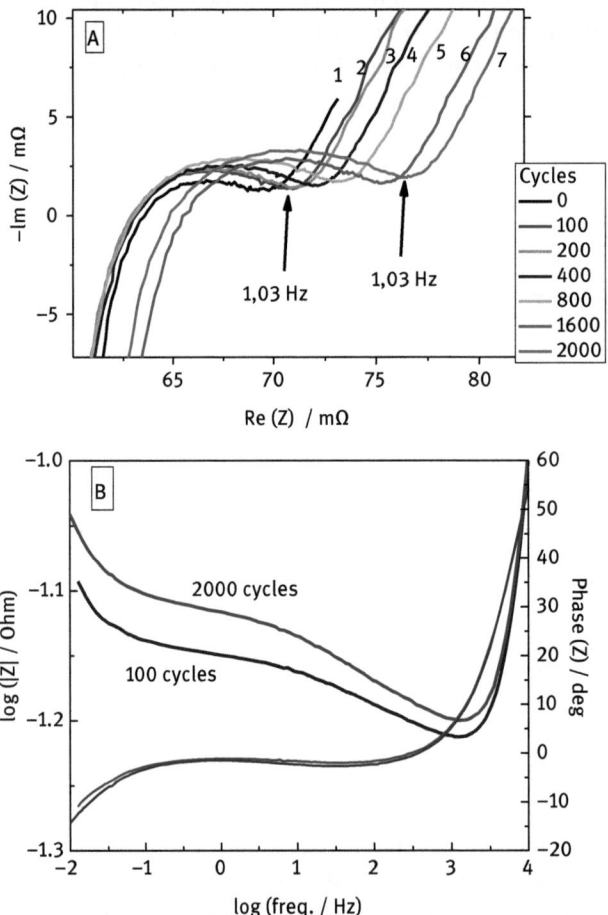

Fig. 7. Typical Nyquist (A) and Bode (B) plot at different SoHs for 1.1 Ah LFP battery.

3.3 Data Modelling

The information provided by the impedance spectroscopy experiments represents the key role for data modelling. To achieve this aim, suitable models are developed, which in the case of LiBs several mechanisms need to be considered. Therefore a classical Randles equivalent circuit was chosen for data interpretation (Fig. 8).

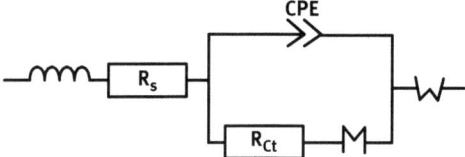

Fig. 8. Proposed equivalent circuit for EIS data modelling.

The proposed model contains inductance, electrolyte and contact resistance (R_s), charge transfer resistance (R_{ct}), constant phase element (CPE), restricted linear diffusion (M) and Warburg element (W). Differential equations describing the electrode behaviour can be derived from the equivalent circuit. The equivalent circuit was chosen based on the very good correlation of the EIS experimental data with the fitting model (Fig. 9).

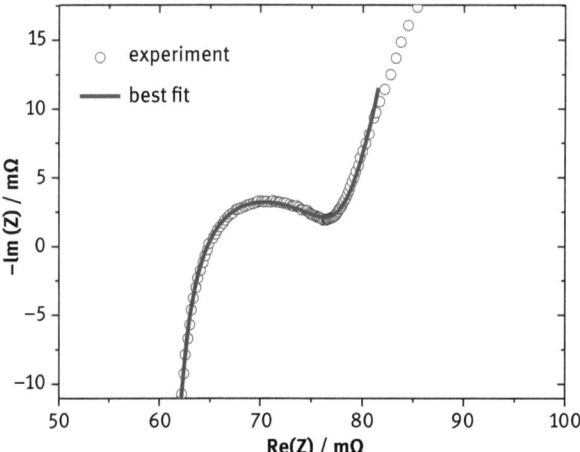

Fig. 9. Correlation of the EIS experimental data with the fitting model. Experimental data are obtained from A123 1.1 Ah LFP battery.

To evaluate the ageing effects it is important to know the expected influence of the equivalent circuit parameters on the shape of impedance spectra. Therefore the parameters are associated with specific frequency ranges, an entire description of the impedance spectra behaviour being obtained. At high-frequency ranges the ohmic, active materials contact and electrolyte resistance is represented by R_s, while R_{ct} describes the charge transfer reaction combined with double-layer capacitance. The Warburg element is responsible for the diffusion reaction of the Li ions into the electrodes and characterizes the impedance spectra at low-frequency ranges. The reason why CPE is used here instead of a capacitance is that the intercalation and de-intercalation of lithium processes are not uniform throughout the surface of the electrodes. Because of the electrode geometry and the connections inside the cell, an inductance was added to the equivalent circuit to take into account the inductive behaviour at high frequencies.

A restricted or blocking diffusion was used due to the complex geometry and morphology of the battery electrode material. The restricted diffusion condition is considered either in a thin film of host material deposited on a substrate impermeable to the diffusing species or in a material foil or in platelet particles. Due to the fact that, the interpretation of the transport phenomena in batteries is in many cases limited by the complex geometry of the electrode and morphology of the material particles, diffusion control (very fast insertion reaction kinetics), linear diffusion and restricted diffusion conditions were assumed.

The parameters of the equivalent circuit were estimated by a non-linear curve-fitting based on a Simplex method for each impedance spectra. The results show that the equivalent circuit parameters vary with the SoC and SoH.

3.3.1 SoC modelling

The inductive distribution of the impedance on the lithium ion battery was attributed to the connectors of the electrode. Therefore, the value of the inductance is considered irrelevant to the battery characteristics and this parameter was omitted during subsequent discussion. Our explanation was supported by previous research results [20, 21], showing that, the inductive behaviour is attributed to the geometrical nature of conductors and not to the faradaic processes in the battery. The EIS spectra were modelled to determine the influence of the impedance parameters as a function of SoC. Fig. 10 shows the evolution of R_s and R_{ct} with the SoC.

Because of the electrolyte resistance and pH decrease upon battery discharge, the bulk series resistance Rs is increasing as the battery becomes progressively more discharged. Regarding the electrolyte concentration, it reaches a conductivity maximum for 100% SoC and therefore increases with discharge. Thus, the change of the electrolyte concentration combined with the contact resistances increases its electric resistance. Furthermore, an increase of charge transfer resistance is observed with decrease of the SoC. This behaviour is attributed to a decrease of exchange current density with decrease of SoC, which reduces the charge transfer reaction at the electrode–electrolyte interface. During discharge or charge the charge transfer reaction takes place from the electrode/electrolyte interface to the interior of the electrode or vice versa. According to these considerations, the R_{ct} increases slightly with the decrease of charge of the battery. Our results are supported by literature data, showing that R_s decreases slowly upon charge and re-increases as SoC reaches 80%. They explain that, the positive reaction products formed upon discharge and the gas dissolution in the electrolyte produced in the overcharge conditions are responsible for the increase of R_s at low and high SoC values [22]. On the other hand, Zhang et al. [23] concluded that, the charge transfer resistance is independent of its SoC, since this electrode does not undergo any significant structural modification during charge–discharge reactions, as these involve only absorption and desorption of hydrogen atoms in the alloy lattice. As conclusion, the small differences measured do not allow accurate determination of SoC from these parameters. The evolution of Warburg element and restricted linear diffusion element with the SoC is presented in Fig. 11.

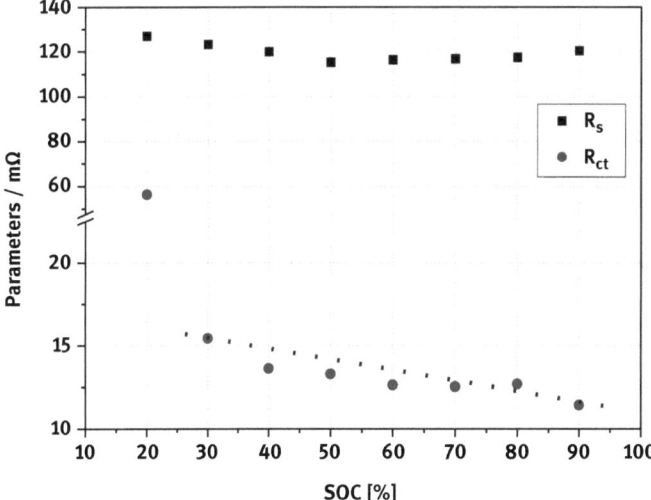

Fig. 10. Evolution of R_s and R_{ct} with the SoC. Experimental data are obtained from A123 1.1 Ah LFP battery.

The Warburg element shows that, the diffusion coefficient (obtained from Warburg) of less-charged batteries is higher than that of fully charged batteries. The higher Warburg values at the lower SoC of the cell would only indicate that longer time periods for the reactants to diffuse through the Nernst diffusion layer would be required. [12]. Another explanation is that the particle radius is kept constant at different SoC levels and any changes are being related to the diffusion coefficient. Thus, the diffusion coefficient of half-charged batteries is higher than of fully charged ones [24]. Moreover the diffusion of ions is limited because of the increased or reduced ion concentration at a specific state of charge. Therefore it is difficult to describe the mass transport effects from electrical circuits having complex equations, many parameters and limited accuracy [2]. Furthermore, because of the separation impossibility of impedance spectra from the anode and cathode, the increase of charge transfer and Warburg element could be probably mainly ascribed to the anode due to the formation of solid electrolyte interface (SEI) layer and decrease of lithium concentration. On the other hand, very low values of the restricted linear diffusion element were observed (Fig. 11). Under restricted diffusion conditions, the current transient caused by a small potential step is not modified, due to the presence of the differential (double-layer) capacitance for times longer than the time constant. Because of the blocking conditions, the provided diffusion element takes on a very low value with respect to the insertion capacitance of the host material.

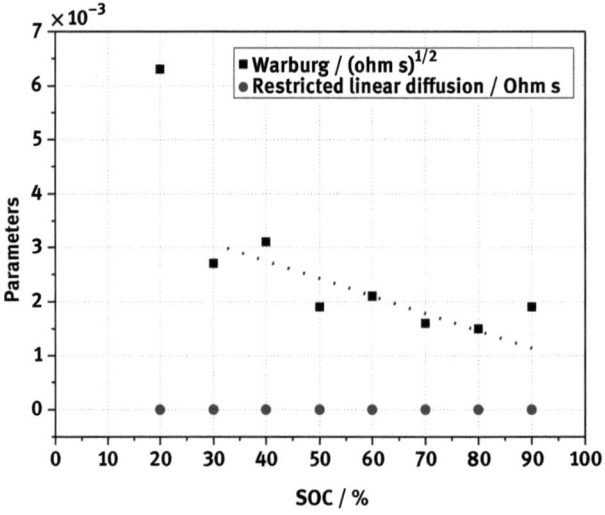

Fig. 11. Warburg element and restricted linear diffusion evolution with the SoC. Experimental data are obtained from A123 1.1 Ah LFP battery.

3.3.2 SoH modelling

As already indicated above (Fig. 7) the impedance spectra change with ageing and represent a good SoH indicator. The dependence of R_s, R_{ct} and W with the cycle number is in Fig. 12 displayed. All the parameters were evaluated for an SoC of 50%.

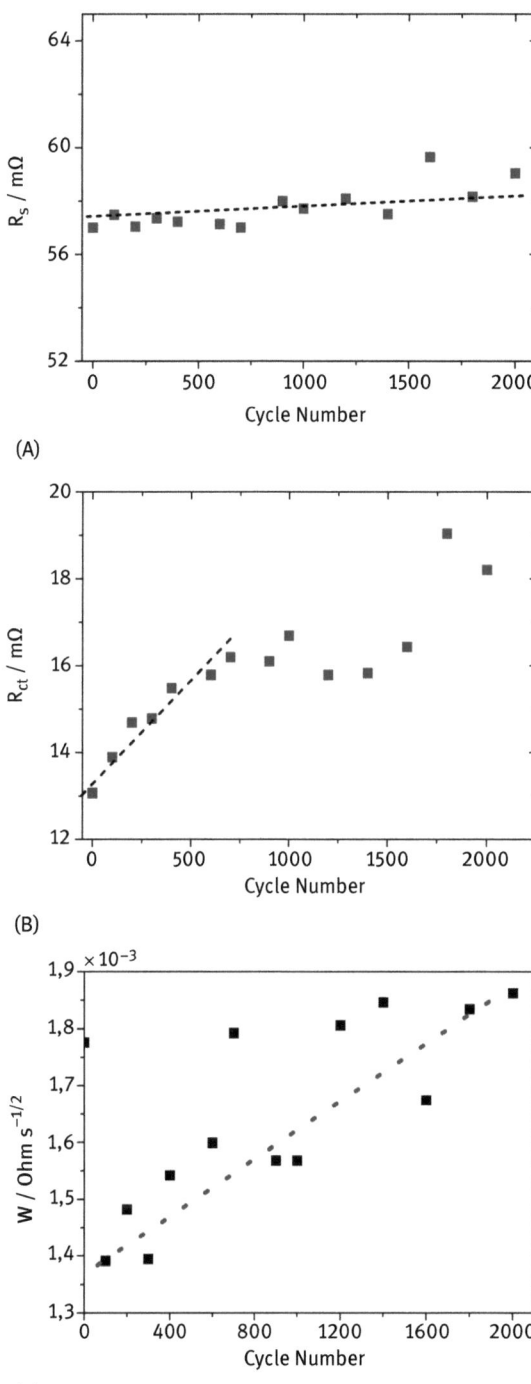

Fig. 12. R_s(A), R_{ct}(B) and W(C) evolution with the SoH. Experimental data are obtained from A123 1.1 Ah LFP battery, SoC 50%.

R_s remains constant within the studied cycle number, showing no ageing of the electrolyte. In particular, after 2,000 cycles of the cell the R_{ct} increases about 50% from the initial value. This result is related to degradation processes taking place in the battery materials and represents an indicator for a decreasing exchange current density resulting from the decreasing lithium concentration and the surface film growth. Another explanation for this obvious increase is that the SEI film growth reduces charge transfer reaction rate at the electrode/electrolyte interface [12]. Because of the instability of electrolyte, the Fe ions are dissolved, resulting in loss of the cathode active material. Furthermore, the Fe ions can reduce on the surface of the graphite anode and catalyze the SEI formation, leading to increase in resistance and loss of Li ions. A notable change was also observed in the Warburg parameter (W) where the 45 angle at the lower frequencies of the graphs would correspond to a semi-infinite diffusion. Deviation from this slope would be due to the effect of porous structure and particle size distribution on the diffusion-related impedance in the active particles [25]. The increasing Warburg element points to a decrease of diffusion coefficient and a decrease of lithium concentration. Moreover, it might be a result of surface film growth. The surface film growth irreversibly consumes lithium and inhibits diffusion [4, 12]. A coarsening effect of the particles was observed by Nagpure et al. [26]. This leads to a decrease in the effective surface area of the particles, which further affects the rate of the reaction. The change in the particle size would also influence the diffusion kinetics of the lithium ions during charging and discharging cycles. Using a classic Randles approach the electrical parameters obtained from impedance data were presented by Tröltzsch et al. [4]. Combining fast ageing tests with the Levenberg–Marquardt method for a better modelling of the impedance spectra, the change of the capacity fade (−14% over 250 cycles), R_{ct} (+62% over 250 cycles) and Warburg element (+72% over 250 cycles) versus cycle number was determined. Finally, the two-electrode cell adopted in this study gives a limitation in differentiating the reaction steps caused by the anode and the cathode. For a better differentiation of the anode and cathode contribution, a three-electrode system will be used and in the future work explained.

4 Conclusions and Outlook

The SoC and SoH of an LFP battery were characterized by EIS and represent the basis for the estimation of a battery capacity. The observed very high sensitivity of model parameters to resistance changes proves that impedance spectroscopy is suitable for characterization of ageing effects. A simplified physical model was chosen to investigate the impedance increase of the cell. It was observed that, the increase is due to the increase of the anode and cathode surface resistances. After 2,000 cycles R_{ct} increases approximately with 50% from the initial value, which indicates a decreasing of the exchange current density. The Warburg element showed a decrease of diffusion coefficient and lithium concentration due to the formation of SEI layer. The charge transfer resistance demonstrates more significant trend with ageing cycles and the results are consistent with experiment-based observations from the literature, which seems to be indicating the potential of the proposed model for battery age estimation. However, the applicability of the proposed model needs to be further investigated. To prove these assumptions and because of the high sensitivity of several model parameters to ageing effects, further investigations such as battery variety, temperature, charge and discharge rate need to be considered. Furthermore, the obtained experimental impedance data will be subjected to simulation to find a suitable and accurate model for describing the SoC and SoH of the battery. For a better understanding of the ageing mechanism EIS data acquisition and modelling of the battery under dynamic conditions (constant current) are as well aimed.

Acknowledgements: The financial support from Europäischer Sozialfonds (ESF) is gratefully acknowledged. The authors are thankful to Pavel Votyakov for measurements assistance.

5 References

[1] J. Vetter, P. Novak, M. Wagner, et al., "Ageing mechanisms in lithium-ion batteries," Journal of Power Sources, vol. 147, pp. 269–291, 2005.
[2] A. Jossen, "Fundamentals of battery dynamics," Journal of Power Sources, vol. 154, pp. 530–538, 2006.
[3] V. Pop, H. Bergveld, P. Notten, et al., "Smart and accurate state-of-charge indication in portable applications," Measurement Science and Technology, vol. 16, pp. R93–R110, 2005.
[4] U. Tröltzsch, O. Kanoun, and H.-R. Tränkler, "Characterizing aging effects of lithium ion batteries by impedance spectroscopy," Electrochimica Acta, vol. 51, pp. 1664–1672, 2006.
[5] R. Mingant, J. Bernard, V. S. Moynot, et al., "EIS measurements for determining the SoC and SoH of Li-ion batteries," ECS Transactions, vol. 33, pp. 41–56, 2011.
[6] N. Omar, M. A. Monem, Y. Firouz, et al., "Lithium iron phosphate based battery – Assessment of the aging parameters and development of cycle life model," Applied Energy, vol. 113, pp. 1575–1585, 2014.
[7] S. Rodrigues, N. Munichandraiah, and K. A. Shukla, "A review of state-of-charge indication of batteries by means of a.c. impedance measurements," Journal of Power Sources, vol. 87, pp. 12–20, 1999.
[8] Z. Yanhui, S. Wenji, L. Shili, et al., "A critical review on state of charge of batteries," Journal of Renewable and Sustainable Energy, vol. 5, p. 021403, 2013.
[9] A. G. Ritchie, "Recent developments and likely advances in lithium rechargeable batteries," Journal of Power Sources, vol. 136, pp. 285–289, 2004.
[10] A. G. Ritchie and W. Howard, "Recent developments and likely advances in lithium-ion batteries," Journal of Power Sources, vol. 162, pp. 809–812, 2006.
[11] A123 Systems. Available at: www.a123systems.com (accessed 5 September 2016).

[12] E. Ferg, C. Rossouw, and P. Layson, "The testing of batteries linked to supercapacitors with electrochemical impedance spectroscopy: a comparison between Li-ion and valve regulated lead acid batteries," Journal of Power Sources, vol. 226, pp. 299–305, 2013.

[13] A. J. Salkind, C. Fennie, P. Singh, et al., "Determination of state-of-charge and state-of-health of batteries by fuzzy logic methodology," Journal of Power Sources, vol. 80, pp. 293–300, 1999.

[14] M. Schönleber, D. Klotz, and E. Ivers-Tiffee, "A method for improving the robustness of linear Kramers-Kronig validity tests," Electrochimica Acta, vol. 131, pp. 20–27, 2014.

[15] M. Dubarry and B. Y. Liaw, "Identify capacity fading mechanism in a commercial $LiFePO_4$ cell," Journal of Power Sources, vol. 194, pp. 541–549, 2009.

[16] M. Dubarry, C. Truchot, and B. Y. Liaw, "Cell degradation in commercial $LiFePO_4$ cells with high-power and high-energy designs," Journal of Power Sources, vol. 258, pp. 408–419, 2014.

[17] Q. Zhang and R. E. White, "Capacity fade analysis of a lithium ion cell," Journal of Power Sources, vol. 179, pp. 785–792, 2008.

[18] A. K. Padhi, K. S. Najundaswamy, and J. B. Goodenough, "Phospho-olivines as positive-electrode materials for rechargeable lithium batteries," Journal of Electrochemical Society, vol. 144, pp. 1188–1194, 1997.

[19] W. J. Zhang, "Structure and performance of LiFePO4 cathode materials: a review," Journal of Power Sources, vol. 196, pp. 2962–2970, 2011.

[20] E. Barsoukov and J. R. Macdonald, Impedance Spectroscopy Theory, Experiment, and Application, 2nd ed. Hoboken, NJ: John Wiley Sons, 2005, p. 91.

[21] S. Thele, M. Buller, R. W. De Doncker, et al., "Impedance-based simulation models of supercapacitors and Li-ion batteries for power electronic applications," IEEE Transactions on Industry Application, vol. 41, pp. 742–774, 2005.

[22] A. Hammouche, E. Karden, and R. W. De Doncker, "Monitoring state-of-charge of Ni–MH and Ni–Cd batteries using impedance spectroscopy," Journal of Power Sources, vol. 127, pp. 105–111, 2004.

[23] W. L. Zhang, M. P. S. Kumar, and S. Srinivasan, "AC impedance studies on metal hydride electrodes," Journal of Electrochemical Society, vol. 142, p. 2935, 1995.

[24] S. E. Li, B. Wang, H. Peng, et al., "An electrochemistry-based impedance model for lithium-ion batteries," Journal of Power Sources, vol. 258, pp. 9–18, 2014.

[25] Y. Zhang, C.-Y. Wang, and X. Tang, "Cycling degradation of an automotive LiFePO4 lithium-ion battery," Journal of Power Sources, vol. 196, pp. 1513–1520, 2011.

[26] S. C. Nagpure, B. Bhushan, S. Babu, et al., "Scanning spreading resistance characterization of aged Li-ion batteries using atomic force microscopy," Scripta Materialia, vol. 60, pp. 933–936, 2009.

Delf Kober, Oliver Görke and Julia Kowal

Ageing Investigation of Lithium Ion LiFePO$_4$ Batteries with a Combination of EIS and Structural Analysis

Abstract: Electrochemical impedance spectroscopy (EIS) measurements and post-mortem analyses are used on commercial LiFePO$_4$ cells to identify the ageing mechanisms and their representation in the impedance spectrum. The cells were cycled up to 1,667 cycles and measured with EIS. At different ageing states two cells were opened and analyzed with X-ray diffraction (XRD) and scanning electron microscope (SEM), one at 0% state of charge (SoC) and one at 100% SoC. First results show slight changes both in the EIS spectra and in the micro-structure with cycling.

Keywords: LiFePO$_4$, ageing, impedance spectroscopy, post-mortem analysis

1 Introduction

Within the scope of energy supply, energy storage applications become more and more important. Since 25 years rechargeable Li ion batteries have been applied successfully in portable electric devices. The understanding of the degradation behaviour of single components and the entire cell is crucial for high-power applications – be mobile or non-mobile – in terms of economic and safety aspects. Despite of the growing interest and the continuous increasing of publications in this field of research, there remain unanswered questions.

One important attribute of a battery is its ageing, which is highly governed by degradation processes. Ageing determines the lifetime of the battery and is caused by both cycling and storage. Among other influences, ambient temperature (as well as electrode potential, depth of discharge (DOD) and cycling rate [1]) affects ageing. The structural changes in the active material (oxidation, corrosion, disordering, micro-cracking), SEI formation, chemical decomposition or dissolution and loss of contact are some of the main reasons for ageing [2, 3]. They lead to loss of accessible capacity and increasing cell impedance. It was found that ageing is more related to the cathode than to the anode. So the cathode is to a high degree responsible for capacity fade [4, 5] and to a certain degree also for an increase of impedance, caused by particle size reduction [5]. The different approaches in literature to model ageing are dealing with the influence of temperature, state of charge (SoC), ΔDOD, U or I(t) and

---

Delf Kober and Oliver Görke, Department of Material Science, Technical University of Berlin, Berlin, Germany
Delf Kober and Julia Kowal, Department of Energy and Automation Technology, Technical University of Berlin, Berlin, Germany

DOI 10.1515/9783110449822-002

cell capacity C [6, 7]. Very few publications deal with the physical causes – outlined above – for ageing models be it the SEI formation [8] or the evolution of electrode porosity [9]. One attempt was done by Zavalis et al. [5] developing a model that describes the electric property resistances and double-layer capacities in respect to surface area of active material and current collector. Diffusion coefficient in active material was considered, too.

The aim of this study is the correlation of the structural parameter with the impedance data from cycled cells (at different cycle numbers) at different ambient temperatures. The structural changes are analyzed by post-mortem analyses of the cycled cells. The cathode and anode are investigated with XRD, SEM as well as impedance spectroscopy with respect to phase changes, particle size and morphology, defect evolution, chemical decomposition and precipitates. Simultaneous impedance measurements give information about the resistances (current collector, electrodes, electrolyte) and double-layer capacitances. The observed ageing phenomena: particle size distribution, phase composition and stresses of the electrodes can be correlated with the results of the EIS measurements in terms of cycle number and ambient temperature.

The results of this study shall contribute to a physical chemical ageing model, which will be implemented in a lifetime prediction model. With the parameterization of the non-destructive EIS technique to the electrode's micro-structure, it is intended to use EIS as diagnostic tool for cells in application and resolve the degradation of the cell components. Supplementary with the capacity fade determined from the cycle curve it is aimed to make predictions to future performances.

2 Experimental

For the study, commercial 1.1 Ah 18650 graphite/LiFePO$_4$ (LFP) cells (manufacturer A123) were used. The test matrix is shown in Fig. 1. The cells were cycled with a Cadex battery testing system at ambient temperature 24 °C between 2 and 3.6 V with 2C. EIS measurements were done in fully discharged and charged states at cycle #1, #112, #333, #667, #1000 and #1333 with a Zahner ZENNIUM Impedance measurement unit under galvanostatic mode with 5 mA amplitude between 10 mHz and 100 kHz. For basic statistics four cells were measured and averaged at each cycle number. The data were fitted with the "Nelder-Mead simplex direct search" algorithm with MATLAB (the used equivalent circuits for the refinement are shown in Fig. 2). After 1,333 and 1,000 cycles two cells were opened, one at 0% SoC and one at 100% SoC. The cathode was separated from anode and separator and washed in dimethyl carbonate (DMC) to remove electrolyte residues. The sample with a diameter of 16 mm was punched out at the centre position both in longitudinal and transversal dimensions of the cathode strip. For micro-structure characterization scanning electron microscopy (SEM) and X-ray diffraction (XRD) were used. The SEM pictures were made from cathodes with a Zeiss Gemini Leo 1540 in topography mode at an acceleration voltage of 3 kV and a magnification of 50.000.

The XRD measurements were done solely on the cathodes with Bruker D8 Advance in Bragg Brentano geometry with copper radiation ($\lambda K \alpha 1: 1.78897 Å$) between 15 and 120 °C 2 θ (step width 0.02 2 θ). The X-ray data were analyzed with Rietveld method with TOPAS software.

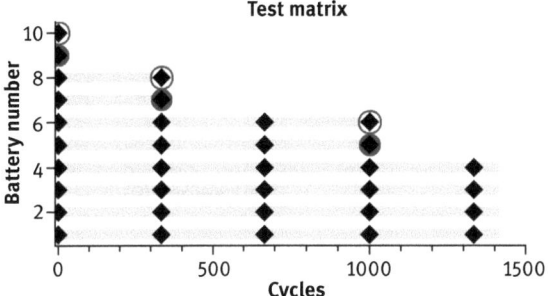

Fig. 1. Test matrix. The gray bars describe the cycling while the diamonds symbolize the EIS measurements at 0% and 100% SoC and the black circles samples taken for the post-mortem analysis at 0% SoC (empty circle) and at 100% SoC (filled circle).

3 Results

3.1 Electrochemical Impedance Spectroscopy

To find the best fit of an equivalent circuit model with the measured data for 0% and 100% SoC, different models were used. At the beginning the physical meaning of the impedance response is not requested. Fig. 2 shows as an example of two Nyquist plots of a cell measured (+) after 300 cycles with fit (o) at 100% SoC (A) and 0% SoC (B). Both curves differ from each other. At 100% SoC the EIS curve consists of an inductive part at higher frequencies described by an inductance L element in series with an ohmic resistance R, describing the shift of 17 mΩ on real impedance axis. Towards lower frequencies there is a flattened semicircle, described with a parallel connection of an ohmic resistance and a constant phase element (CPE). The flattened semicircle can be interpreted as a composition of several semicircles (RC elements). The resistance contributes to the charge transfer reaction and the capacity to the double-layer properties. Each of the two electrodes has a characteristic charge transfer resistance and double-layer capacity. At low frequencies the data describe a tail, which in turn is described with a CPE in series. The equivalent circuit and the influenced curve section is shown in Fig. 2A. This tail represents the contribution of the solid state diffusion in the electrodes to the impedance response. Even if a diffusion element would be more appropriate to explain the physical effect, we found a CPE element as appropriate for an adequate fit.

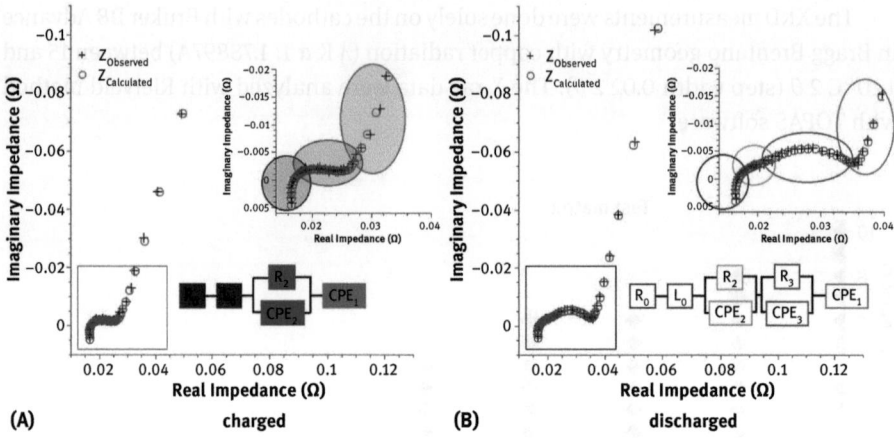

Fig. 2. Nyquist plots of a charged (A) and discharged (B) cell measured (+) after 333 cycles and fit (o). The inlets are zooms from the framed region. The influence of the single elements of the equivalent circuit (inlet at the bottom) on the plot is marked (coloured bubbles) in the inlets

At 0% SoC (Fig. 2B) a second distinguishable semicircle is visible, which is described by a second parallel connected ohmic resistance with CPE. It could be deduced that the charge transfer and double-layer capacity of one electrode is dependent from SoC. There are ongoing investigations for a clarification of the development of the semicircles over SoC aimed to find one appropriate model over the entire SoC range. The equivalent circuit is also shown. As it can be seen in Fig. 2 the chosen model fits the measured data quite well.

One important indicator for ageing is fade of capacity. Fig. 3 shows the development of the charge capacity over cycling. After 1,333 cycles, the capacity has decreased up to 5%. The fitted parameters for R1, QCPE2 and αCPE2 are shown in Fig. 4 (colour code see Fig. 2). Because of the existence of an additional RC element at 0% SoC, there are three more parameters to fit. The other elements, whether charged (filled symbols) or discharged (empty symbols), are in same range. After 1,333 cycles the parameters show only slight trends. With continued cycling, larger changes are expected.

Parameter R1 increases at 0 and 100% SoC with cycling. This could be related to the growing SEI layer. With increasing layer thickness it becomes harder for lithium ions to pass through the layer and consequently the resistance increases. A closer look at CPE2 shows contrary tendencies. While at 0% SoC Q is increasing with cycling, the opposite is true at 100% SoC. The exponent α is decreasing at 0% SoC and increasing at 100% SoC. The bigger errors at 100% SoC result from difficulty of getting smooth EIS spectra. The data at 100% SoC were scattering stronger. The change of the CPE parameter at 0% SoC could be explained with a surface roughening of both electrodes. At 100% SoC the low start value and the following increase of α could be related to the chosen model of only one R-CPE element as discussed above. The contribution of two RC elements fitted with one R-CPE element leads to the flattened semicircle explaining the low value of 0.4.

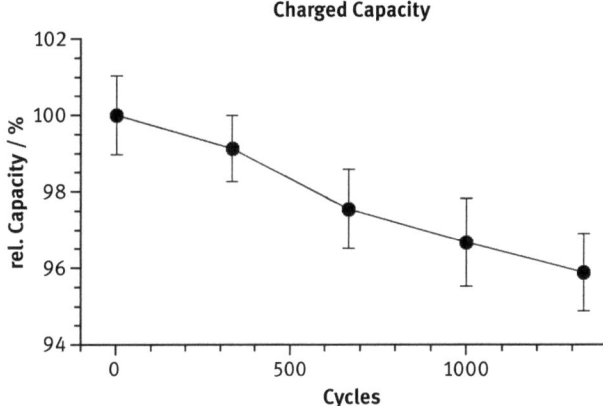

Fig. 3. Averaged charge capacity over cycle number. For averaging four cells were used. Error bars show deviation within the four samples. The averaged capacity value at one cycle was normalized to 100%.

3.2 Post-mortem Analysis

The SEM images of the anodes at 100% SoC (Fig. 5A–C) show an amorphous film, blurring the contours. Because of the electronic conductivity of the carbon matrix in both electrodes, there was no additional carbon coating for the SEM measurements. The bright blurs at 100% SoC indicate an electrostatic charge of the surface, that is the layer has insulating properties. This film could be the SEI layer. At one cycle there is a haze over the entire image leading to poor contrast, while with increasing cycling number the contrast is increasing and the electrostatic charges are locally distributed and more severe, leading to the assumptions of a thicker film growth. Cracks are visible in the film at all ageing states at 100% SoC, which could be explained rather with uncoiling the battery for preparing it for post-mortem analysis than for ageing. At 0% SoC (Fig. 5D–F) where no Lithium is in the anode, no such film is visible. It is not clarified yet if it is the result of preparing the SEM samples under air atmosphere or an inherent battery characteristic. For the SEM images of the cathodes (Fig. 5G–I) there are no visible changes in morphology after 1,000 cycles. For a quantified analysis of the electrode roughening, laser-microscopic investigations and measurements after Brunauer–Emmett–Teller (BET) are remaining. To clarify the chemical composition and identify the different particles in the images, energy-dispersive X-ray spectroscopy

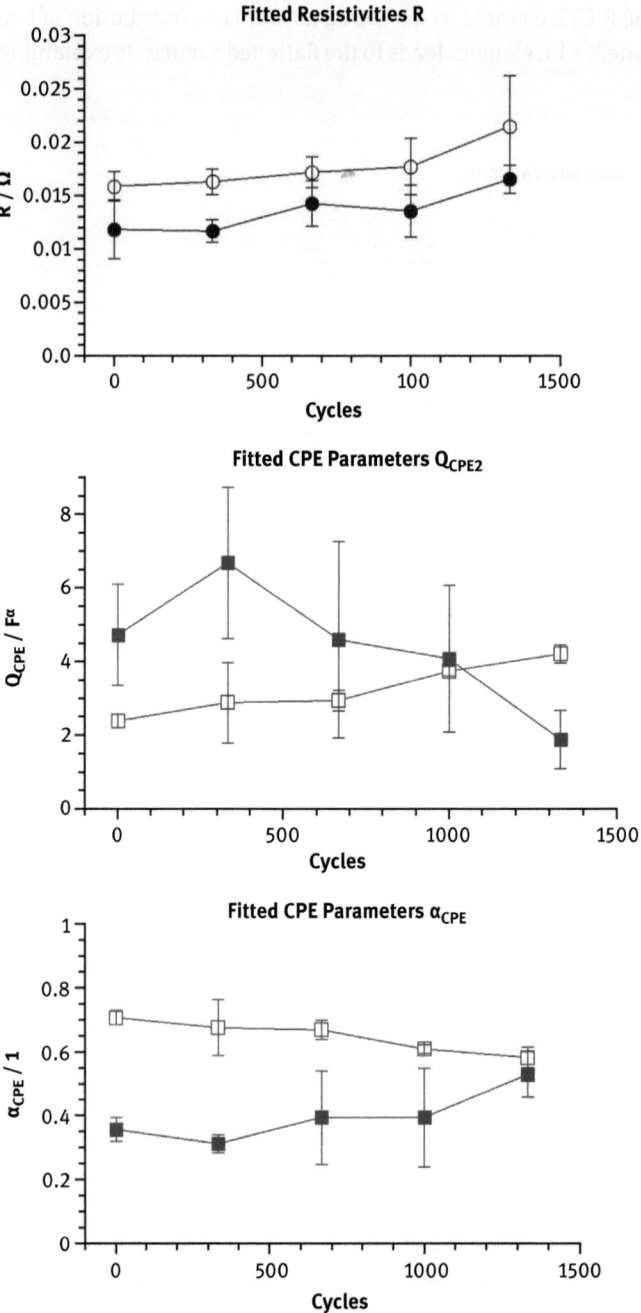

Fig. 4. Refined resistance R1 (top) and QCPE2 (middle) and α CPE2 (bottom) over cycles from charged (filled) and discharged state (empty). The coloured symbols correspond to the elements in the equivalent circuits in Fig. 2.

(EDX) measurements and inductively coupled plasma optical emission spectrometry (ICP-OES) have still to be done.

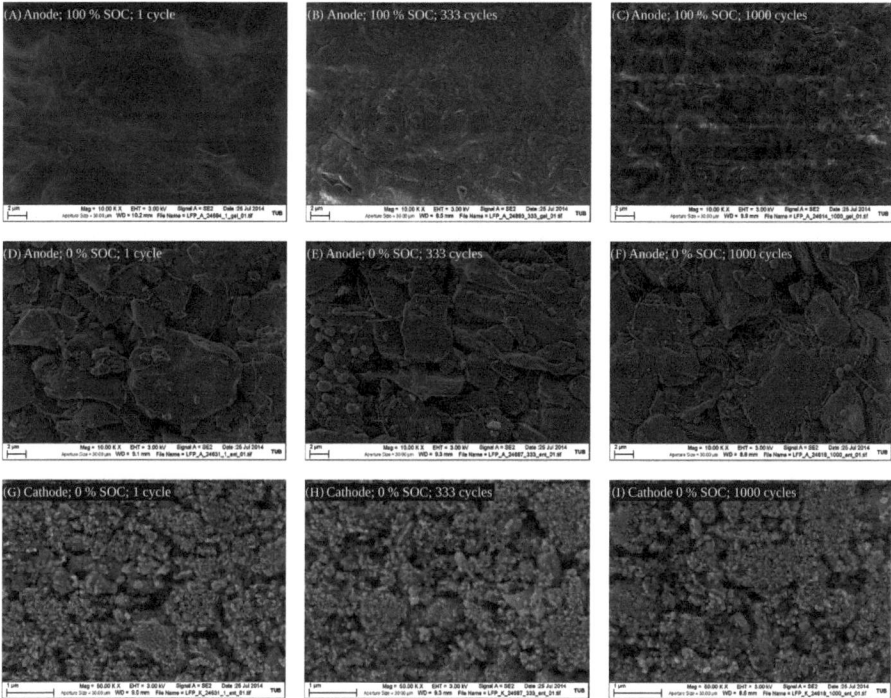

Fig. 5. SEM images of (A) to (C): anodes at 100% SoC; (D) to (F): anode at 0% SoC and (G) to (I): cathode at 0% SoC over cycle #1 (left). #333 (middle) and #1,000 (right). The images of the cathodes at 100% SoC were also made but are skipped here because of no significant differences to 0% SoC.

X-ray diffraction is a suitable technique for the investigation of phase changes during de-/lithiation in the electrodes, especially in LiFePO$_4$ with its two phases and structure-related ageing effects like strain, crystallite size and lattice occupancies. Fig. 6 shows the diffraction patterns of a cell at 0% SoC and one at 100% SoC after 333 cycles. The ageing state was arbitrarily chosen only for demonstration purpose. The analysis of the XRD data of the cathodes shows two phases: a lithium-poor phase Li0.05FePO$_4$ and a lithium-rich phase Li0.89FePO$_4$ (after Yamada et al. 10), depending on the SoC. At 0% SoC Li0.89FePO$_4$ is the predominant phase, while at 100% SoC it is Li0.05FePO$_4$. For the Rietveld refinement these two phases were used as model structures plus aluminium because of the penetration of the X-rays into the current collector. The weighted reliability factor (Rwp) and the goodness of fit (GoF) of the refinement were 16 and 1.16, respectively. The focus here is on the refined parameters of crystallite size and occupancy of Fe atoms on the Fe lattice position. For the charged and discharged states only the predominant phase with its characteristics is shown: Li0.89FePO4 at 0% SoC and Li0.05FePO4 at 100% SoC. In Fig. 7A the development of the crystallite size of both LiFePO4 phases over the ageing state is shown. For both SoCs and so for both phases the crystallite size is increasing with increasing ageing. The bigger the crystallites are the longer time need the lithium ions to diffuse into the entire volume. If this longer time is not given due to high c-rate cycling, not all lithium ions participate to the cycling process leading to lower capacity. This corresponds to the slight capacity decrease in Fig. 3. Another aspect of higher crystallite sizes is grain coarsening. This observation corresponds with the EIS results, especially with the CPE2 correlation at 0% SoC. The development of the Fe occupancy over the cycling number is shown in Fig. 7B. For both phases a decrease of Fe on the Fe lattice positions is visible. Since the Fe ions work as redox center, missing iron leads to less consumable Lithium in the LiFePO4 structure and this in turn causes capacity fade. Where the missing iron is located could not be clarified at this stage. For clarification reasons ICP-OES and EDX measurements on the anode are planned. Micro-structure changes in the cathode due to ageing could be found, although there is no visible change in the particle morphology indicated with SEM.

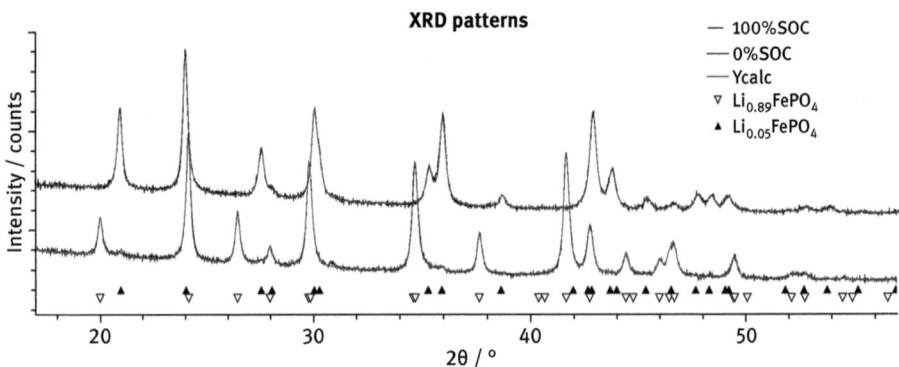

Fig. 6. Two exemplary XRD patterns of two cells after 333 cycles; at 0% SoC (gray line) and at 100% SoC (black line). The triangles mark the positions of atomic planes of Li0.05FePO4 phase (filled) and of the Li0.89FePO4 phase (empty triangles). The shown patterns are only a section. The full measurement range was between 15 and 120 2θ, which is necessary for the Rietveld analysis (gray line). For the refinement the two LiFePO4 phases were used plus an aluminum phase, which appears at higher angles, is caused by the interaction of the X-rays with the current collector.

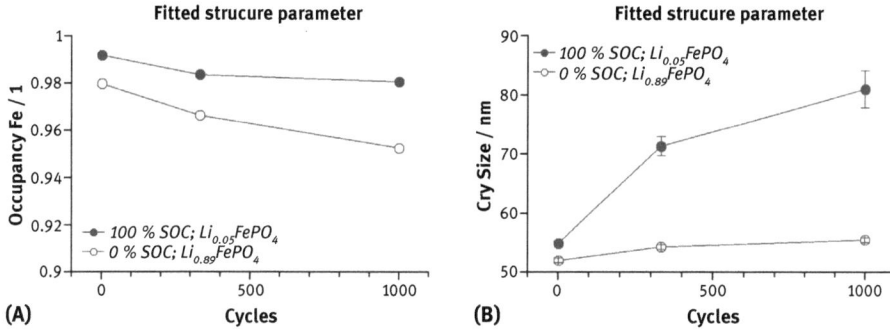

Fig. 7. Results of the Rietveld analysis. The crystallite size (A) and the Fe occupancy on Fe lattice positions (B) at 0% SoC (empty circle) and at 100% SoC (filled circles) over cycle number. The parameters at 0% SoC correspond to the Li0.89FePO4 phase because of its dominance at this SoC, while at 100% SoC Li0.05FePO4 is dominating and that is why its parameters are shown here.

4 Summary and outlook

Commercial LiFePO$_4$ cells were cycle aged. After 1,333 cycles the capacity decreased up to 5%. First EIS measurements show slight changes in the impedance parameter. Further tests with up to 2,000 cycles are still remaining. Parallel to the EIS measurements the cells will be opened and analyzed post-mortem with respect to surface morphology, phase composition, crystallite size and Fe occupancy with XRD and SEM. First results could be correlated to the SEI layer, grain coarsening and active material loss.

5 References

[1] J. Veter, P. Novák, M. Wagner, et al., "Ageing mechanisms in lithium-ion batteries," Journal of Power Sources, vol. 147, pp. 269–281, 2005.
[2] P. Arora, R. White, and M. Doyle, "Capacity fade mechanisms and side reactions in lithium-ion batteries," Journal of the Electrochemical Society, vol. 145, pp. 3647–3667, 1998.
[3] J. Fergus, "Recent developments in cathode materials for lithium ion batteries," Journal of Power Sources, vol. 195, pp. 939–954, 2010.
[4] J. Zhou and P. Notten, "Studies on the degradation of Li-ion batteries by the use of microreference electrodes," Journal of Power Sources, vol. 177, pp. 553–560, 2008.
[5] T. G. Zavalis, M. Klett, M. H. Kjell, et al., "Aging in lithium-ion batteries: model and experimental investigation of harvested LiFePO$_4$ and mesocarbon microbead graphite electrodes," Electrochimica Acta, vol. 110, pp. 335–348, 2013.
[6] M. Safari and C. Delacourt, "Simulation-based analysis of aging phenomena in a commercial graphite/LiFePO$_4$ cell," Journal of the Electrochemical Society, vol. 158, pp. A1436–A1447, 2011.

[7] M. Safari and C. Delacourt, "Modeling of a commercial graphite/LiFePO$_4$ cell," Journal of the Electrochemical Society, vol. 158, pp. A562–A571, 2011.
[8] M. Broussely, S. Herreyre, P. Biensan, et al., "Aging mechanism in Li ion cells and calendar life predictions," Journal of Power Sources, vol. 97–98, pp. 13–21, 2001.
[9] G. Sikha, B. Popov, and R. White, "Effect of porosity on the capacity fade of a lithium-ion battery – Theory," Journal of the Electrochemical Society, vol. 151, pp. A1104–A1114, 2004.
[10] A. Yamada, H. Koizumi, S. Nishimura, et al., "Room-temperature miscibility gap in LiFePO$_4$," Nature Materials, vol. 5, pp. 357–360, 2006.

Paul Büschel, Thomas Günther and Olfa Kanoun
Streamlining Calculation of the Distribution of Relaxation Times from Time Domain Data

Abstract: For the identification of the linear transfer behaviour of lithium ion batteries, impedance spectroscopy is widely used. The data evaluation using models however requires some knowledge for the model building and fitting. One method to support this evaluation is the calculation of the distribution of relaxation times based on the transformation of impedance data. This calculation can be difficult because of the limited number of data points and noise in the data. In this chapter a novel method is presented, which uses the time-domain data directly without calculating the impedance first. Additionally, the method features an iterative scheme with very little computational effort. With no need to store the whole data set, it is attractive for embedded solutions. The algorithm is explained and the results obtained are compared with impedance measurements.

Keywords: Impedance, distribution of relaxation times, time domain, streamlining

1 Introduction

The identification of the linear behaviour of lithium ion batteries is usually done by impedance spectroscopy. Often the impedance is evaluated using models formed by electrical equivalent circuits. To build such a model *a-priori* knowledge on the ongoing mechanisms is needed.

One way to support the model building and subsequent parameterization is to use the concept of distribution of relaxation times (DRT) [1]. Here the capacitive part of the impedance is represented by a Voigt model, which consists of a series connection of RC elements, each characterized by a time constant τ and a weight $y(\tau)$. The impedance of such a circuit can be described by an integral equation combining the polarization resistance R_{pol} with the density $y(\tau)$:

$$Z(\omega) = R_0 + R_{pol} \int_0^\infty \frac{1}{1+j\omega\tau} y(\tau)\,d\tau. \tag{1}$$

Separation into real and imaginary part and subsequent discretization of the above integral equation leads to a linear system of equations:

$$\mathbf{y} = \mathbf{Ax} \tag{2}$$

Paul Büschel, Thomas Günther and Olfa Kanoun, Chair for Measurement and Sensor Technology, Technische Universität Chemnitz, Chemnitz, Germany

DOI 10.1515/9783110449822-003

with $y \in R^{m \times 1}$ being the measurements at m frequency points, $A \in R^{m \times n}$ is the coefficient matrix and $x \in R^{n \times 1}$ the desired DRT at n support points in the time constant domain. Solving for **x** is an ill-posed inverse problem for ratios $\frac{m}{n} < 4$ [2]. The solution process for this linear system must be stabilized against amplification of noise in the solution. This is usually done by incorporation of direct regularization [3] or by implementing relaxed iterative solution algorithms, which also implement regularization [4–6]. The choice of a suitable regularization itself is an optimization problem, which is hard to solve. This is the major drawback in this impedance data analysis approach. In the resulting DRT spectrum ($y(\tau)$ vs. τ) one can separate effects by their corresponding peaks. This can be used for model building: the number of mechanisms can be identified and their corresponding time constants can be fixed during the complex non-linear least squares fitting. Additionally the DRT allows for simplified visual inspection of impedance measurement data as well as for further automated analysis of the DRT peaks.

2 Novel Algorithm

The DRT can be used to perform a time domain simulation of the voltage response of a battery to an applied current signal. In [7] the goal is to minimize the number of RC elements needed for simulation. The novel algorithm will make use of the simulation approach to perform a streamlining calculation of the DRT based on time domain data. For the simulation a set of RC elements is described in the time domain by z-transform of their transfer function. Additional separation of the weighting $y(\tau)$ from the temporal behaviour leads to a recursive formulation for low pass filtering the current signal with time constants defined by the DRT support points τ and a subsequent dot product for the weighting step to obtain the voltage response:

$$y_k = \alpha x_k + (1 - \alpha) y_{k-1} \tag{3}$$

with x being the current at time step k, α is the filter coefficient for time constant τ and the voltage response is defined by $u_k = \langle y_k, y(tau) \rangle$. The digital filters for the time constants of the DRT form a filter bank. No computational burden results from this as it requires only the last output to be stored for the next step. The weighting forms the second step to obtain the dynamic voltage response. If the open circuit potential and the iR voltage drop are added, one obtains the overall response of the battery. Writing the above for subsequent time steps results in a linear system:

$$\begin{vmatrix} \vdots \\ u_{sim,k} \\ \vdots \end{vmatrix} = \begin{vmatrix} \vdots & \vdots & \vdots & & \vdots \\ 1 & i_{sim,k} & y_{\tau_1,k} & \cdots & y_{\tau_n,k} \\ \vdots & \vdots & \vdots & & \vdots \end{vmatrix} \cdot \begin{bmatrix} U_0 \\ R_0 \\ y(\tau_1) \\ \vdots \\ y(\tau_n) \end{bmatrix}$$

$$\mathbf{y} = \mathbf{A}\mathbf{x}. \tag{4}$$

This linear system has virtually unlimited rows. The calculation can be done for an infinite duration as long as the system state does not change.

For the calculation of $\mathbf{x}$, the DRT, this needs to be inverted. Under the requirement of persistence of excitation, two possible ways exist. First one can store current and voltage of the battery for a certain amount of time, calculate the coefficient matrix and directly solve for the unknowns:

$$\mathbf{x} = (\mathbf{A}^T\mathbf{A})^{-1}\mathbf{A}^T\mathbf{y}. \tag{5}$$

This imposes computational burden because of the wide frequency range that needs to be observed. Typically in the range of 1 kHz down-to 10 mHz, sampling frequencies of at least 5 kHz and a signal duration of around 300 s for three periods of the lowest signal frequency are required. This leads to 1.5×10^6 samples. Together with a suitable number of time constants to be identified, the inversion of the resulting matrix $\mathbf{A}$ is computationally very expensive. The second approach for solving the linear system is an iterative calculation of the unknowns. Best known for this is the recursive least squares (RLS) algorithm [8]. It updates the solution of $\mathbf{x}_k$ at time step k based on the current row a_k of the matrix $\mathbf{A}$ and the previous solution of $\mathbf{x}_{k-1}$. To reduce the computational cost and the influence of the inverted information matrix incorporated in the RLS algorithm, the Kaczmarz algorithm is used [6]. By combining the iterative calculation of the matrix row a_k by use of a filter bank and the solution update, the algorithm is reduced to an iterative calculation scheme using only voltage and current at time step k as well as the last solution $\mathbf{x}_{k-1}$:

$$\mathbf{x}_k = \mathbf{x}_{k-1} + \lambda \frac{u_{meas} - \langle a_k, \mathbf{x} \rangle}{\|a_k\|^2} a_k. \tag{6}$$

Instead of using the simulated, the measured voltage is now used together with the corresponding rows of $\mathbf{A}$ to update the solution. A relaxation parameter $\lambda < 1$ ensures a stable solution. The overall algorithm is shown in Fig. 1. It is a two-step iteration of updating the filter bank with a subsequent Kaczmarz update step of the solution. The streamlining nature of the algorithm is clearly visible.

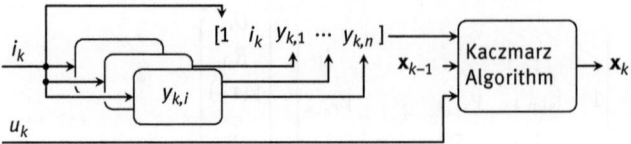

Fig. 1. Novel algorithm for the DRT calculation from time domain data.

3 Experimental

The algorithm was tested with data obtained for battery measurements. The same current/voltage data set was used in all calculations. A multi-spectral excitation was chosen consisting of 101 logarithmically spaced frequencies in the range of 1 kHz down-to 10 mHz with random phase in $[0, 2\pi]$. The overall excitation signal was normalized to have an RMS value of 0.5 A. Sampling frequency was 5 KHz. This signal was applied to a 5-Ah Lithium polymer cell that was used for testing. Measurements were performed using a NI PCI4461 Dynamic Signal Analyzer card together with a Servowatt DCP 520/30-based current source. For comparison the measured impedance was compared with the impedance calculated from the DRT, which was obtained with the new algorithm described above. This way it was avoided to compare two solutions of two ill-posed problems (DRT calculated from time-domain and from impedance data). In contrast the calculation of the impedance from time-domain data as well as from the DRT is well posed. Fig. 2 shows the impedance measured at 10% state of charge (SoC) steps. The change in the impedance semicircle can be identified on a visual basis, but it is hard to identify the number of mechanisms that lead to the impedance, which is necessary for further model-based data evaluation. To be able to investigate the impedance semicircle more in detail, the DRT was calculated from the same data using the algorithm described above. The result is given in Fig. 3. One can see that multiple processes are present in the frequency range covered by the electrode

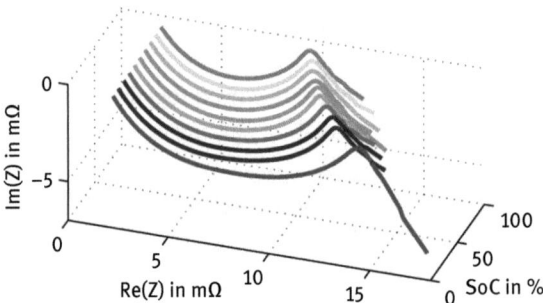

Fig. 2. IS spectra measured with multisine signal.

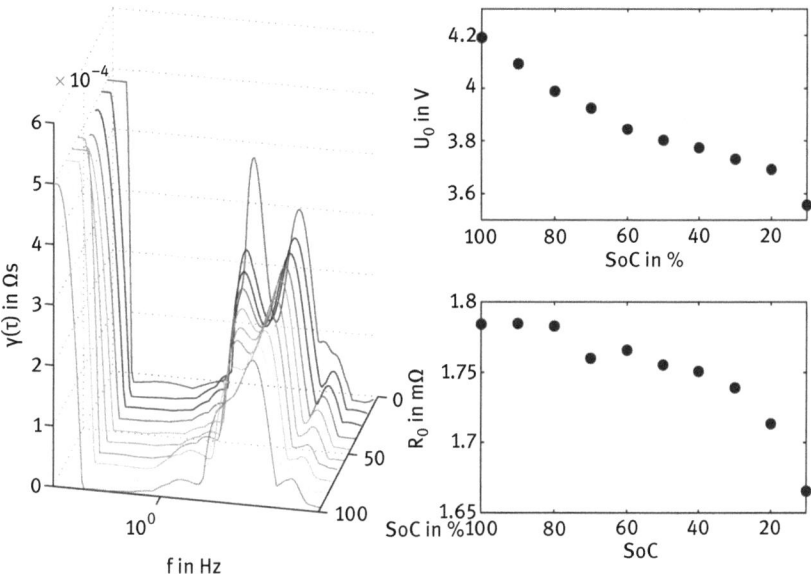

Fig. 3. DRT obtained by the novel algorithm. Frequency f is used instead of τ for better comparison to the impedance data.

semicircle. The size and position of the peaks slightly change with SoC. Additionally the open circuit voltage and the internal resistance are shown. As comparison, Fig. 4 shows the impedance spectra calculated from the DRT as well as the one obtained via Fourier transform as overlay. Both spectra are in good agreement, especially in the

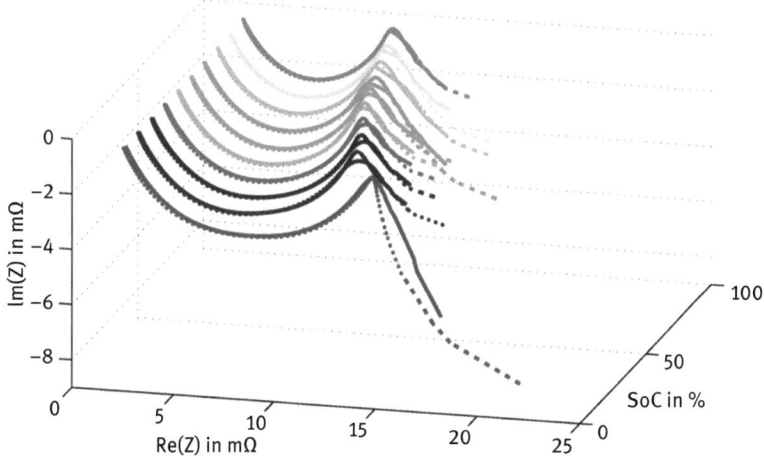

Fig. 4. Comparison of IS spectra obtained the classical way (solid) and via DRT (dashed).

semicircle range. The low-frequency part of the spectra shows some disturbance for the DRT-based calculation. This is due to the short measurement time of only three periods for the lowest frequency.

4 Conclusion and Outlook

A new algorithm for the calculation of the DRT from time domain data was presented. The results indicate a good performance. The DRT is calculated with high accuracy, especially in the frequency range characteristic for electrode processes. The use of time domain data for the calculation of the DRT creates a highly over-determined system of equations which, together with the iterative solution update strategy, converges to the true solution. This way of calculating the DRT requires no optimization of a regularization parameter, which, in contrast, is necessary for the classic impedance-based calculation approach. The algorithm itself is of streamlining nature and of low complexity. It has low-memory requirements and is thus highly suitable to be implemented in an embedded measurement system.

5 References

[1] E. Barsoukov and J. R. Macdonald eds, Impedance Spectroscopy, Theory, Experiment and Applications, 2nd ed. Hoboken, NJ: John Wiley Sons, Inc., 2005.
[2] M. Schonleber, D. Klotz, and E. Ivers-Tiffee, "A method for improving the robustness of linear Kramers-Kronig validity tests," Electrochimica Acta, vol. 131, pp. 20–27, 2014.
[3] A. Tikhonov, A. Goncharski, V. Stepanov, et al., Numerical Methods for the Solution of Ill-posed Problems, Boston, MA: Kluwer Academic Publishers, 1995.
[4] L. Landweber, "An iteration formula for Fredholm integral equations of the first kind," American Journal of Mathematics, vol. 73, no. 3, pp. 615–624, 1951.
[5] G. Cimmino and C. N. delle Ricerche, "Calcolo approssimato per le soluzioni dei sistemi di equazioni lineari," Tech. Rep., Istituto per le applicazioni del calcolo, 1938.
[6] S. Kaczmarz, "Angenaherte Auflosung von Systemen linearer Gleichungen," Bulletin International de l'Academie Polonaise des Sciences et des Lettres SÃľrie B, vol. 35, pp. 355–357, 1937.
[7] J. P. Schmidt, P. Berg, M. Schonleber, et al., "The distribution of relaxation times as basis for generalized time-domain models for li-ion batteries," Journal of Power Sources, vol. 221, pp. 70–77, 2013.
[8] M. Werner, Digitale Signalverarbeitung mit MATLAB®-Praktikum: Zustandsraumdarstellung, Lattice, 2008.

Christian Reinke, Kristian Nikolowski, Mareike Wolter and Alexander Michaelis

Influence of the Anode Graphite Particle Size on the SEI Film Formation in Lithium-Ion Cells

Abstract: Electrochemical impedance spectroscopy (EIS) is used as a tool to investigate the formation process of a lithium-ion cell. The usability of EIS was demonstrated for two anode active materials with different particle sizes. The initial charging of the cell was interrupted when a defined anode-half cell potential (vs. Li/Li+) was reached, in order to measure an impedance spectrum. This was fitted with a common equivalent circuit model and the SEI film resistance (RSEI) was extracted. Results show that for both active materials RSEI initially increases until an anode half-cell potential of approximately 0.2 V is reached. Subsequently the RSEI shows a sharp decline for both active materials. The SEI film resistance is significantly higher for smaller particles, indicating that a less conductive SEI is built on smaller particles during formation.

Keywords: Lithium-ion cells, electrochemical impedance spectroscopy, SEI formation

1 Introduction

Expanding life time as well as lowering production costs of lithium-ion cells are two important subjects, which are addressed in the current research of lithium-ion batteries. The first charge/discharge cycles (referred to as the formation) have a major contribution to the production costs, since high investments are required to provide a sufficient amount of formation equipment. Furthermore, since the process takes up to 3 weeks with different aging and charging steps [1], a lot of stock capacity is necessary.

During the formation process a so-called solid–electrolyte interface (SEI) film is built on the active sites of graphite anodes. This film consists of solid decomposition products of the electrolyte, which is unstable in the working potential of a graphite-based anode (below 1.2 V vs. Li/Li+). The SEI film is present on any active graphite sites in contact with the electrolyte [2]. The anode half-cell potential of a fresh assembled lithium-ion cell is about 3.0 V versus Li/Li+ and decreases when the cell formation process is initiated. In the voltage range >1.2 V the surface groups of the graphite are reduced. In between 1.2 and 0.2 V most of the SEI film is built, mainly by reducing the electrolyte solvents and lithium is not significantly intercalated into the graphite particles yet. Finally, below 0.2 V, lithium is intercalating into the graphite particles besides some charge losses due to further SEI film formation [3, 4].

Christian Reinke and Alexander Michaelis, Institut für werkstoffwissenschaft, Technische Universität Dresden, Dresden, Germany
Kristian Nikolowski, Mareike Wolter and Alexander Michaelis, Fraunhofer IKTS, Dresden, Germany

Since the SEI film is an electrical insulator (but a lithium-ion conductor), the electrolyte is protected from further decomposition. The SEI film formation process is crucial in terms of aging, performance and safety [5–7], when using a graphite anode in lithium-ion cells. Therefore the formation is mostly executed under controlled conditions by the cell manufacturer. The SEI film consists of several organic (e.g. $CH_2OCO_2Li)_2$) and inorganic (e.g. Li_2CO_3) components. To understand the building mechanism, investigations on the formation process are mostly done by analyses of the SEI film using extensive (in terms of experimental effort) and destructive methods, which give direct information about the composition and quality of the SEI film (Fourier transform infrared spectroscopy (FTIR), X-ray photoelectron spectroscopy (XPS), scanning electron microscope (SEM)) [2]. In most of these studies the SEI film was analyzed after several cycles and not during the actual formation process.

Electrochemical impedance spectroscopy (EIS) is a widely used tool in lithium-ion battery research for non-destructive studies during operation. For example EIS has successfully been used to investigate the aging behaviour of lithium-ion cells [7, 8] and was implemented in battery management systems (BMSs) to determine the state (e.g. state of charge, state of health, internal temperature) of lithium-ion cells [9–11]. EIS has furthermore been used as a non-destructive method to evaluate the SEI formation behaviour [4, 12–14].

Regarding the formation process, little is known about the interaction between anode material properties and process parameters of electrode manufacturing on one side, and the formation process and cell performance on the other side. Focus of this study is to investigate the influence of different particle sizes of the anode active materials on the SEI film formation in lithium-ion cells. To characterize this influence during operation, EIS is used.

2 Experimental

All the electrodes used for this study were produced on a pilot plant for electrode fabrication. Two commercially available anode active materials with different particle sizes were used (see Tab. 1). Anodes were produced by casting an aqueous solution of graphite as active material, CMC/SBR binder and carbon black as conductive additive on a copper foil. For the cathodes, Li(Ni$_{0.33}$Co$_{0.33}$Mn$_{0.33}$)O$_2$ (Toda NM3100) was used as active material, PVDF as binder and carbon black as conductive additive. The NMP-based cathode slurry was casted on an aluminium foil. The electrodes were calendared to a porosity of 30%. For all measurements a 1-M solution of LiPF$_6$ in EC:DEC (1:1) (LP40, BASF) with 1% vinylen carbonate (Sigma–Aldrich) as additive was used as electrolyte. A polyolefin separator (PP-PE-PP trilayer 2325, Celgard) was used.

The three-electrode pouch cells (see Fig. 1) were prepared as described elsewhere [15, 16].

Tab. 1. Properties of the investigated anodes

Anode ID	A1	A2
Graphite manufacturer, product name	Timcal, Timrex SLP6	Timcal, Timrex SLP10
d_{90}	6.5 μm	12 μm
A_{BET}	14 $m^2 g^{-1}$	9.5 $m^2 g^{-1}$
Weight ratio, binder	4.5 % CMC 2 % SBR	
Weight ratio, conductive additive	2 % carbon black	
Mass loading	5.3 mg cm^{-2}	5.4 mg cm^{-2}

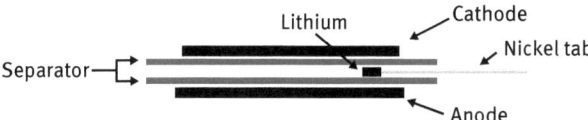

Fig. 1. Schematic cross section of the three-electrode pouch-bag cell configuration used in this study.

As reference electrode a piece of lithium foil was attached to a nickel tab. A uniform thickness of the lithium reference electrode was realized by roll pressing. The reference electrode was sandwiched between two separators. Subsequently, the components were sealed with an aluminium laminate film package. After sealing, leaving one side open, the pouch cells were dried at 60° for 16 h under vacuum and transferred into an argon-filled glove box. Inside the glove box, they were filled with electrolyte. The last sealed seam was done under vacuum. The anode area of the cell was 21.15 cm^2, while the cathode area was 25 cm^2. Furthermore, capacity tests were done in coin cells with an active area of 1.3 cm^2 and a non-woven polypropylene separator (Viledon FS 2190, Freudenberg) to determine the irreversible charge losses during formation. Therefore both anodes were cycled with lithium as counter electrode. Capacity tests were executed with C/10 between 1.2 V and 10 mV.

The three-electrode pouch cells were charged with C/10 until a defined anode half-cell potential was reached (0.8, 0.4, 0.2, 0.1 and 0.025 V vs. Li/Li+). After applying an open-circuit voltage (OCV) period of 30 min, an impedance spectrum was measured. Subsequently the constant current phase was continued until the next anode half-cell potential was reached (see Fig. 2).

The cell impedance was measured by applying an AC bias of 0.7 mA amplitude over the frequency range of 100 kHz to 100 mHz. The impedance spectra were modelled with an equivalent circuit containing a resistor in series with two RC circuits and a Warburg impedance element (see Fig. 3) [17]. Constant phase elements (CPEs) were used instead of ideal capacitors due to the fractal nature of the electrode/electrolyte interface [18].

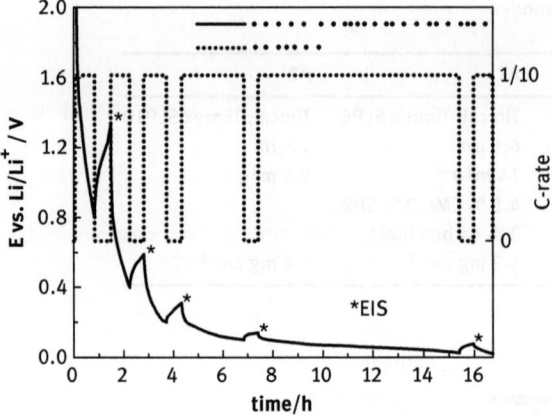

Fig. 2. Anode half-cell potential and c-rate vs. time. Experimental set-up for EIS investigations (EIS was executed at *).

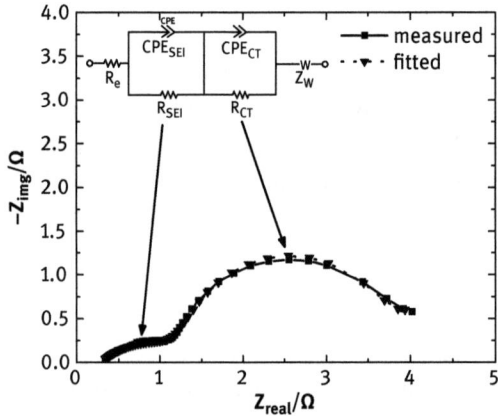

Fig. 3. Measured and fitted Nyquist plot for the A2 electrodes at 0.5 V anode half-cell potential vs. Li/Li+. For data fitting the shown equivalent circuit was used.

The SEI film properties correspond to the high-frequency RC circuit, where the resistance RSEI is due to the lithium ion diffusion through the SEI film (ionic resistance) and the capacitance CPESEI is accorded to the dielectric nature of the SEI film [19]. Modelling was executed with the software Gamry Echem Analyst.

All experiments were carried out in a climate chamber (CTS GmbH) at 30° with a multi-channel potentiostat/galvanostat (BaSyTec GmbH) in combination with a frequency response analyzer (Interface 1000, Gamry).

3 Results

The charge that is consumed until an anode half-cell potential of 0.2 V during the initial half-cycle is generally attributed to the SEI film formation. The intercalation of lithium ions in the graphite is initiated at an anode half-cell of 0.2 V. Therefore the irreversible capacity was divided in two regions (see Fig. 4):
- above 0.2 V: $Q_{(irr,>0.2V)}$ is defined as the charge that is consumed until an anode half-cell potential of 0.2 V vs. Li/Li+ is reached during the initial charging. Approximately all charge is used to build the SEI film
- below 0.2 V: beside intercalation of lithium in graphite, some charge is still used for SEI film formation ($Q_{(irr,>0.2V)}$)

Fig. 5 shows the irreversible capacity for both anodes separated for the two regions (above 0.2 V and below, respectively). The irreversible capacity is higher for smaller particles in both regions (A1 in Fig. 5). This higher irreversible capacity is due to the fact that smaller particles have an higher active surface area in contact with the electrolyte and therefore more charge is required for SEI film formation. The irreversible capacity seems to be a linear function of the BET surface area [20, 21]. About 2/3 of the irreversible capacity is consumed above and 1/3 below 0.2 V, which is similar to literature findings for different material systems [4]. In summary, it was shown in agreement with literature that a smaller graphite particle size leads to

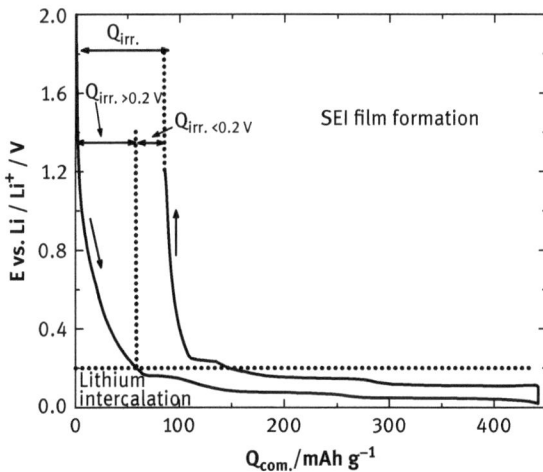

Fig. 4. Anode half-cell potential vs. the cumulated specific capacity (intercalation denoted as positive and deintercalation as negative. $Q_{1stint.}$ and $Q_{1stdeint.}$ are the capacities of the initial charging step and the initial discharging step, respectively.

significantly higher irreversible capacity during initial charging, due to the higher active surface area of the particles.

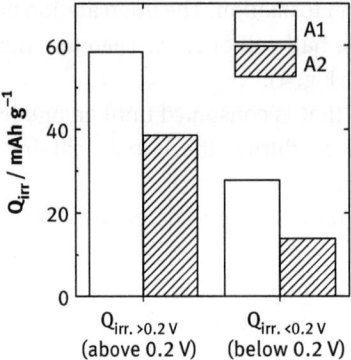

Fig. 5. Irreversible capacity in the first intercalation half cycle above and below 0.2 V anode half-cell potential vs. Li/Li+.

To obtain information not only about the quantity of the SEI film (indicated by the irreversible capacity) but also about the quality of the SEI film, RSEI has been extracted from the measured impedance spectra as described in the experimental part. The obtained values for RSEI are normalized to the active surface (see ABET in Tab. 1) and plotted as function of the voltage at the end of each OCV period (see Fig. 6). For both anodes, RSEI initially increases with decreasing voltage and reaches a maximum of 78 kΩ cm^2 at about 0.3 V for A1 and of 16 kΩ cm^2 at 0.5 V for A2, respectively. RSEI for A1 (smaller particles) is about 4.8 times higher in the voltage range above 0.2 V compared with A2. In the lithium-intercalation region (0.2–0.025 V) A1 shows a sharper decline compared with A2. However, after full intercalation of lithium into the graphite particles, the SEI film resistance is still 3.2 times higher for the smaller particles.

It is proposed in literature that the SEI film formed above 0.2 V is loose and highly resistive [4]. When the lithium intercalation in the graphite particles is initiated (in other words, when the anode half-cell potential drops below 0.2 V), lithium ions migrate through this loose SEI film and are used for restructuring reactions to build a compact and conductive SEI film [4, 12, 22]. This is supported by the quantitative analysis of the charge measurements, since there is still a significant amount of irreversible charge losses in the potential range below 0.2 V (see Fig. 5).

When assuming that the properties of the SEI film are similar on both graphite types, one would expect that RSEI is equal for both particle sizes during the SEI film formation process, when normalizing it to the active surface area. However, the results of this study show that, especially above an anode half-cell potential of 0.2 V, significantly different normalized SEI film resistances are measured. This means that the SEI films must differ in composition, thickness and/or structure. This could be due to the slower SEI film growth rate (in terms of thickness increase per time unit) on smaller particles when using a constant current. As a consequence a different composition and/or structure would be formed, which would lead to a different ionic resistance of the SEI film. Furthermore, transport properties in the electrolyte could be influenced by the higher particle density (in terms of particles per volume unit) when using smaller particles. This would lead to a lack of educts for the SEI film formation reaction, which would also lead to a different SEI film quality in structure and/or composition. During the intercalation of lithium into the graphite, the RSEI declines for both graphite types, which must be due to restructuring of the SEI film. As can be seen in Fig. 6, the decline is significantly higher for smaller particles, indicating that the restructuring reactions must differ for both graphite types.

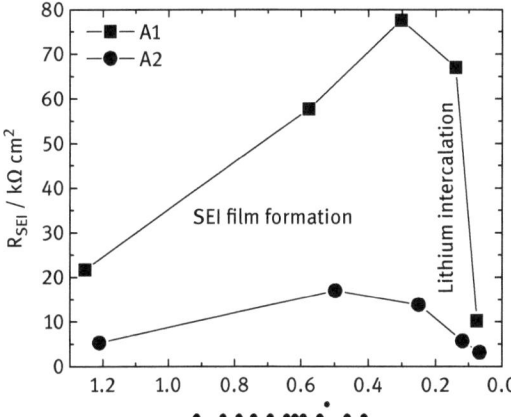

Fig. 6. SEI film resistance as a function of anode half-cell potential vs. Li/Li+ for the first intercalation half-cycle for both anode materials.

4 Conclusion

To optimize the formation process it is essential to understand how active material properties as well as electrode manufacturing parameters influence the SEI film formation and, eventually, the cell performance. In this study EIS was used as a non-destructive tool to investigate the formation process. Therefore the RSEI was extracted from anode half-cell impedance spectra obtained from a three-electrode cell configuration by fitting it to an adequate equivalent circuit model. This RSEI can be interpreted in terms of SEI film formation progress and morphological changes of the SEI film.

In this study two anode active materials with different particle sizes were investigated. The results show that for the anode active material with smaller particle size the SEI film resistance is significantly higher than for larger particles. This effect is pronounced, especially in the region where only SEI film formation (and no lithium intercalation into graphite) occurs. In the potential range below 0.2 V, the decline in RSEI is sharper, indicating that the SEI film restructuring reactions, which take place when lithium is migrating through the SEI film are different for both active materials.

To clarify if the proposed mechanisms influence the SEI film morphology as assumed, further experiments will be performed in which the influence of the current profile and the temperature on the SEI film resistance are studied. Additionally, information of the development of the SEI film composition and thickness during formation would be useful.

5 References

[1] D. L. Wood, J. Li, and C. Daniel, "Prospects for reducing the processing cost of lithium ion batteries," Journal of Power Sources, vol. 275, pp. 234–242, 2015.

[2] P. Verma, P. Maire, and P. Novák, "A review of the features and analyses of the solid electrolyte interphase in Li-ion batteries," Electrochimica Acta, vol. 55, no. 22, pp. 6332–6341, 2010.

[3] P. Novák, F. Joho, M. Lanz, et al., "The complex electrochemistry of graphite electrodes in lithium-ion batteries," Journal of Power Sources, vol. 97–98, pp. 39–46, 2001.

[4] S. Zhang, M. S. Ding, K. Xu, et al., "Understanding solid electrolyte interface film formation on graphite electrodes," Electrochemical and Solid-State Letters, vol. 4, no. 12, p. A206, 2001.

[5] V. A. Agubra and J. W. Fergus, "Lithium ion battery anode aging mechanisms," Materials, vol. 6, pp. 1310–1325, 2013.

[6] A. Barré, B. Deguilhem, S. Grolleau, et al., "A review on lithium-ion battery ageing mechanisms and estimations for automotive applications," Journal of Power Sources, vol. 241, pp. 680–689, 2013.

[7] J. Vetter, P. Novák, M. R. Wagner, et al., "Ageing mechanisms in lithium-ion batteries," Journal of Power Sources, vol. 147, no. 1–2, pp. 269–281, 2005.

[8] U. Tröltzsch, O. Kanoun, and H.-R. Tränkler, "Characterizing aging effects of lithium ion batteries by impedance spectroscopy," Electrochimica Acta, vol. 51, no. 8–9, pp. 1664–1672, 2006.

[9] P. L. Moss, G. Au, E. J. Plichta, et al., "An electrical circuit for modeling the dynamic response of Li-ion polymer batteries," Journal of the Electrochemical Society, vol. 155, no. 12, p. A986, 2008.

[10] T. K. Dong, A. Kirchev, F. Mattera, et al., "Dynamic modeling of Li-ion batteries using an equivalent electrical circuit," Journal of the Electrochemical Society, vol. 158, no. 3, p. A326, 2011.

[11] C. Fleischer, W. Waag, H.-M. Heyn, et al., "On-line adaptive battery impedance parameter and state estimation considering physical principles in reduced order equivalent circuit battery models," Journal of Power Sources, vol. 260, pp. 276–291, 2014.

[12] S. Zhang, K. Xu, and T. Jow, "EIS study on the formation of solid electrolyte interface in Li-ion battery," Electrochimica Acta, vol. 51, no. 8–9, pp. 1636–1640, 2006.

[13] M. Itagaki, K. Honda, Y. Hoshi, et al., "In-situ EIS to determine impedance spectra of lithium-ion rechargeable batteries during charge and discharge cycle," Journal of Electroanalytical Chemistry, vol. 737, pp. 78–84. 2014.
[14] M. Itagaki, N. Kobari, S. Yotsuda, et al., "In situ electrochemical impedance spectroscopy to investigate negative electrode of lithium-ion rechargeable batteries," Journal of Power Sources, vol. 135, no. 1–2, pp. 255–261, 2004.
[15] M. Dollé, F. Orsini, A. S. Gozdz, et al., "Development of reliable three-electrode impedance measurements in plastic Li-ion batteries," Journal of the Electrochemical Society, vol. 148, no. 8, p. A851, 2001.
[16] E. Barsoukov, "Kinetics of lithium intercalation into carbon anodes: in situ impedance investigation of thickness and potential dependence," Solid State Ionics, vol. 116, no. 3–4, pp. 249–261, 1999.
[17] D. Andre, M. Meiler, K. Steiner, et al., "Characterization of high-power lithium-ion batteries by electrochemical impedance spectroscopy. II: Modelling," Journal of Power Sources, vol. 196, no. 12, pp. 5349–5356, 2011.
[18] E. Barsoukov and J. R. Macdonald, eds, Impedance Spectroscopy: Theory, Experiment, and Applications, 2nd ed. Hoboken, NJ: Wiley, 2005.
[19] E. Barsoukov, "Effect of low-temperature conditions on passive layer growth on Li intercalation materials," Journal of the Electrochemical Society, vol. 145, no. 8, p. 2711, 1998.
[20] M. Winter, "Graphites for lithium-ion cells: the correlation of the first-cycle charge loss with the Brunauer-Emmett-Teller surface area," Journal of the Electrochemical Society, vol. 145, no. 2, p. 428, 1998.
[21] F. Joho, B. Rykhart, A. Blome, et al., "Relation between surface properties, pore structure and first-cycle charge loss of graphite as negative electrode in lithium-ion batteries," Journal of Power Sources, vol. 97–98, pp. 78–82, 2001.
[22] C. R. Yang, J. Y. Song, Y. Y. Wang, et al., "Impedance spectroscopic study for the initiation of passive film on carbon electrodes in lithium ion batteries," Journal of Applied Electrochemistry, vol. 30, no. 1, pp. 29–34, 2000.

Thomas Günther and Olfa Kanoun
Frequency-Dependent Phase Correction for Impedance Measurements

Abstract: Impedance spectroscopy is a widely used method for characterization of materials and devices in R&D. The integration of this method into portable applications like battery management systems is challenging because of the limitations arising from target applications. Here the measurement time, computational load and the required memory are highly critical aspects. In this chapter it is shown how synchronous sampling can reduce the complexity of the measurement process in multi-channel measurement systems in terms of hardware complexity, total number of measurements and thereby the computational load and memory consumption.

Keywords: Impedance spectroscopy, embedded system, digital signal processing

1 Introduction

Impedance Spectroscopy is widely used for system identification [1, 2]. Therefore measurement techniques such as gain-phase detectors or a 'simple' sampling of the perturbation and reaction in the time domain is commonly used [3]. For embedded systems the hardware complexity is a critical aspect in the design process. The measurement method should get along with as less as possible peripheral components and computational power. Sampling of data in the time domain together with a successive transformation in the frequency domain using Goertzel-Filters [4] is a simple and flexible approach [3].

In Fig. 1 the generalized timing for the sampling-based measurement process in multi-channel impedance measurement systems is depicted. f_s is the sampling rate and Δ represents a delay between the samples of the different measurement channels. The alignment of the measurement with respect to the cycle time of the sampling process is represented by t_0. For simultaneous sampling $\Delta = 0$, for $\Delta \neq 0$ the sampling is synchronous. Anyhow the Nyquist criterion needs to be fulfilled to avoid artefacts caused by aliasing in the measurement results.

Following it will be shown why synchronous sampling can be beneficial compared with simultaneous sampling, especially in the case of multi-channel measurement systems.

Thomas Günther and Olfa Kanoun, Chair for Measurement and Sensor Technology, Technische Universität Chemnitz, Chemnitz, Germany

DOI 10.1515/9783110449822-005

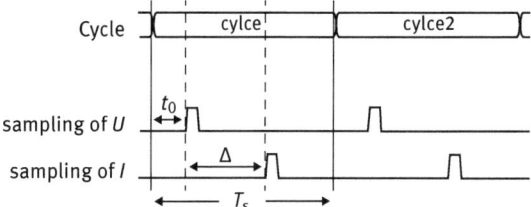

Fig. 1. Alignment of AD conversion during continuous sampling at a sampling rate of $f_s = 1/T_s$.

2 Method

2.1 Measurement Model

For impedance measurement, that is on batteries using the sampling technique, the perturbation as the current flow as well as the reaction as a voltage drop is measured. Whether the current or the voltage is the perturbation usually depends on the characteristics of the device under test. For low ohmic devices a current perturbation is preferred, for high ohmic devices a voltage perturbation in most cases.

In Fig. 2 the time-dependent current-flow i and voltage-drop u are depicted as two different measurement channels. The voltage drop has a second dependency besides the time caused by the shift Δ of sampling points between both measurement channels (compare with Fig. 1). The device under test (DUT) itself is assumed to be linear, frequency dependent and constant in its characteristics during the measurement.

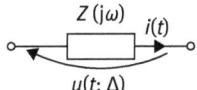

Fig. 2. Measurement model of complex impedance Z with the current i and voltage u.

The current applied to the DUT and the resulting voltage drop are formulated as in the equations 1 and 2. φ_i and φ_u indicate an absolute phase shift with respect to a reference mark in time, see t_0 in Fig. 1. The phase angles $\varphi_{\epsilon,i,u}$ indicate a correction term caused by uncertainties in the timing like jitter.

$$i(t) = I_0(\omega) \cdot e^{j(\omega t + \varphi_i + \varphi_{\epsilon,i})} \qquad (1)$$

A correction term φ_Δ for the delay between the measurement of voltage and current is introduced for the voltage channel.

$$u(t,\Delta) = U_0(\omega) \cdot e^{j(\omega t + \phi_u + \varphi_{e,u} + \varphi_\Delta)} \tag{2}$$

From the previous equations the impedance Z is represented by the expression 3

$$Z(t,\Delta) = \frac{u(t,\Delta)}{i(t)} \tag{3}$$

After reformulation of expression 3 using the equations 1 and 2 and grouping-related parts the following equation is obtained.

$$Z(\omega,\Delta) = \frac{U_0(\omega)}{I_0(\omega)} \cdot e^{j(\phi_u - \phi_i)} \cdot e^{j\varphi_\Delta} \cdot e^{j(\varphi_{e,u} - \varphi_{e,i})} \tag{4}$$

This equation contains four parts from left to right:
- Magnitude of the DUTs impedance, for the frequency ω
- Phase angle of the DUTs impedance, as from measurements using simultaneous sampling of perturbation and reaction
- Phase error caused by the delay Δ between the samples in the measurement channels from synchronous sampling
- Phase noise, that is due to clock jitter

Errors due to phase noise will be assumed to be negligible as the sampling rate (5 kHz) is magnitudes smaller compared with the clock of some megahertz being used for triggering of ADCs.

2.2 Phase Correction

In the continuous sampling process the delay between perturbation and reaction is constant, compared with Fig. 1. According to the shift property of the Fourier transform, this results in a frequency-dependent phase shift of the Fourier coefficient X [5].

$$\mathcal{F}(x(t-\Delta)) = X(\omega) \exp^{-j\omega\Delta} \tag{5}$$

Thereby compensation of the phase error in the impedance Z can be done using the correction term in equation 6.

$$\varphi_\Delta = \omega\Delta \tag{6}$$

3 Experimental Validation

A lithium battery perturbed using a multi-sine sequence was used for verification of the described approach. The multi-sine signals contained equally distributed spectral components on a logarithmic frequency axis from 1 kHz down to 10 mHz at a sampling rate of 5 kHz. The impedance spectrum of the three scenarios, simultaneous measurement with $\Delta = 0$ (Z_{ref}), synchronous measurement with $\Delta \neq 0$ ($Z_{shifted}$) and the compensated impedance spectrum Z_{comp} using the described approach, is depicted in Fig. 3.

The compensated impedance as well as the reference impedance show a good match. In Figs 4 and 5 the magnitude and phase errors after compensation compared with reference method using simultaneous sampling are shown.

The deviation for both the relative error of phase and magnitude after phase compensation is about 0.1% or less. Consequentially synchronous sampled data instead of simultaneous sampled data can be used for the measurement of impedance spectra.

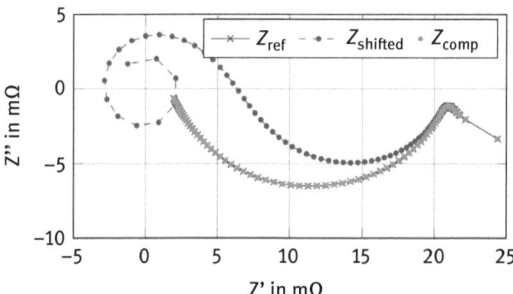

Fig. 3. Impedance Z_{ref} using classical measurement with simultaneous sampling, synchronous sampling and phase-corrected impedance from synchronous sampled data.

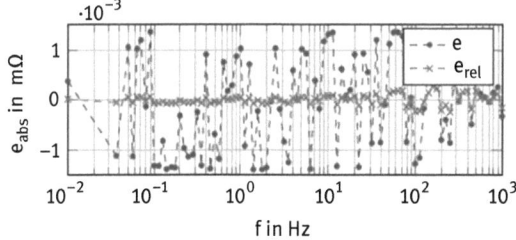

Fig. 4. Magnitude error of the impedance Z_{comp} with respect to the impedance Z_{ref} after error compensation induced by the delay Δ in the measurement.

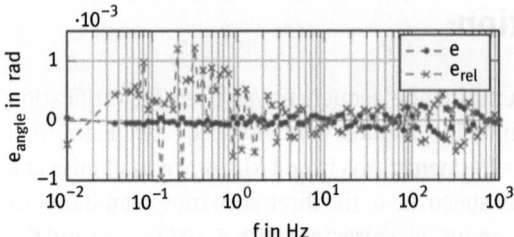

Fig. 5. Phase error of the impedance Z_{comp} with respect to the impedance Z_{ref} after error compensation induced by the delay Δ in the measurement.

4 Conclusion

By using synchronous instead of simultaneous sampled data the following degrees of freedom get available.
- Reduction of hardware complexity by using one ADC for multiple measurement channels with successive sampled data within one cycle
- Reduction of the amount of data to be measured and processed in multi-channel measurement systems measuring the perturbation once and the reaction of multiple channels
- The method applies for all types of perturbation commonly used in measurements for IS like single- and multi-sines, chirps, steps, …

5 References

[1] E. Barsoukov and J. R. Macdonald, Impedance Spectroscopy: Theory, Experiment, and Applications, 2nd ed. Hoboken, NJ: John Wiley Sons, 2005, 616 pp.
[2] M. Orazem and B. Tribollet, Electrochemical Impedance Spectroscopy, ser. The ECS Series of Texts and Monographs. Hoboken, NJ: Wiley, 2011.
[3] T. Günther, P. Büschel, and O. Kanoun, "Eingebettetes impedanzmesssystem für das batteriemanagement in elektrofahrzeugen," Tm - Tech. Mess., vol. 81, no. 11, pp. 560–565, 2014.
[4] G. Goertzel, "An algorithm for the evaluation of finite trigonometric series," American Mathematical Monthly, 1958.
[5] I. N. Bronstein, H. Mühlig, G. Musiol, et al., Taschenbuch der Mathematik, 9th ed. Haan-Gruiten: Europa-Lehrmittel, 2013, 1280 pp.

Sebastian Socher, Claudius Jehle and Ulrich Potthoff

On-line State Estimation of Automotive Batteries using In-situ Impedance Spectroscopy

Abstract: Safe operation of Li-ion traction batteries in electric cars is one of the main requirements for a broad acceptance of this technology. Safety critical factors such as over-temperature conditions must be reliably monitored and captured. Increasingly strict safety regulations in the automotive sector imply rising challenges to both car manufacturers and battery system suppliers and hence resilient, reliable yet inexpensive tools for battery state diagnosis are necessary to address these topics. The approach presented here is a battery diagnosis tool based on in-situ galvanostatic electrochemical impedance spectroscopy, which can be integrated into a standard battery management system. The algorithms, which are used to convert the impedance measurements into valuable battery temperature estimations, are based on numeric correlations between the impedance and these state variables. It is shown that the method can be used even in non-equilibrium state of the battery, thus improving the functional safety in electric cars during the usage of a battery pack by redundant monitoring of the battery temperature.

Keywords: Electric cars, battery temperature, battery safety, impedance spectroscopy

1 Introduction

State-of-the-art battery monitoring in electric vehicles mainly consists of battery voltage, direct current and temperature measurements [1]. The interpretation of the measured data depends on the algorithms, which are implemented in the battery management system. There exists a variety of algorithms, which differ in their kind of data analysis, accuracy and required computational power.

In recent years efforts have been made to determine battery states like temperature, state of health or state of charge by means of electrochemical impedance spectroscopy. Numerous solutions for battery temperature estimation were presented using correlations between cell temperature and single-frequency impedance measurement data [2, 3] or broad-frequency band impedance measurement data [4–6]. Similar methods were also used for state-of-health estimation [7, 8] and state-of-charge estimation [9]. It could be shown that these techniques lead to reasonable results on laboratory scale. However, the implementation of hardware

---

Sebastian Socher, Claudius Jehle and Ulrich Potthoff, Fraunhofer Institute for Transportation and Infrastructure Systems IVI, Dresden, Germany

DOI 10.1515/9783110449822-006

and appropriate algorithms into automotive battery management systems, which are needed for impedance measurements, is still a challenge due to the ambitious functional safety requirements of automotive industry. Furthermore, the cited approaches are all based on impedance measurements performed while the batteries were in equilibrium state meaning that no direct current (DC) was applied during or before the impedance measurements, but especially a temperature estimation is useful just during current loads to monitor the increase of the battery temperature.

The approach presented here is a battery diagnosis tool based on in-situ galvanostatic electrochemical impedance spectroscopy, which can be embedded into a standard battery management system. It is designed for on-line battery state estimation. In this study, it is shown that impedance measurements can be successfully applied for battery temperature estimation during charge and discharge processes as well as dynamic current loads.

2 Experimental Details

In this study the impedance of 21 Ah Li(NMC)O_2 automotive-grade pouch cells were characterized. Regarding the data sheet these cells can be charged with up to 3 °C between 0 °C and 55 °C and discharged with up to 5 C between −20 °C and 55 °C. Impedance measurements were performed in galvanostatic mode with an excitation amplitude of 200 mA in a frequency range from 10 kHz to 10 mHz using a Gamry IF1000 galvanostat. To investigate the influence of a DC-offset on the impedance measurements a DC-offset of 0.5 A was superimposed on the AC-excitation in charge as well as in discharge direction. The ambient temperature was adjusted from 0 °C to 50 °C using a Binder MK53 temperature chamber.

3 Results and Discussion

Estimating the battery temperature via in-situ impedance measurements requires a correlation between both values being unique for each cell. Getting this numerical correlation out of impedance measurements in equilibrium state of the respective battery can be achieved easily (see Chapter 1). For practical applications detailed information about the dependence of this correlation from the DC-load and the DC-profile is necessary to monitor the temperature also during these non-equilibrium states.

3.1 Impedance Measurements

Fig. 1 shows the experimental sequence of the impedance measurements. The reference measurement (1) was carried out over the full frequency range and under

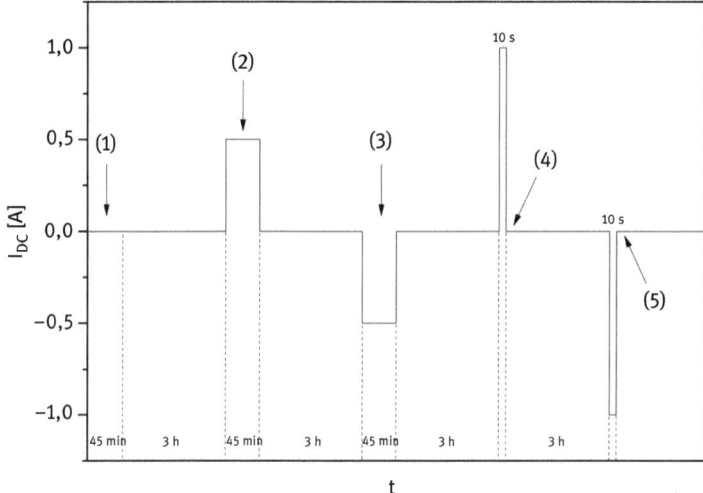

Fig. 1. Experimental procedure for battery characterization: (1) $I_{DC} = 0$ A; (2) $I_{DC} = 0.5$ A; (3) $I_{DC} = -0.5$ A; (4) $I_{DC} = 0$ A after 1 A pulse and (5) $I_{DC} = 0$ A after -1 A pulse.

electrochemical and thermal equilibrium conditions meaning that the DC-offset was $I_{DC} = 0$ A. Non-equilibrium measurements were performed over the full frequency range with a DC-offset of $I_{DC} = 0.5$ A (2) and $I_{DC} = -0.5$ A (3). Each measurement took about 35 min. The state-of-charge variation between the beginning and the end of both measurements was about 1.5% meaning that the expected change in the impedance spectra between the equilibrium and non-equilibrium measurements could rather be attributed to the DC-offset than to the state of charge dependency of the impedance itself. Measurements (4) and (5) are single-frequency impedance measurements determining the impedance of the cells right after a current pulse of 1 A and -1 A for 10 s, respectively. After each measurement the cells were stored for 3 h to achieve electrochemical and thermal equilibrium again.

Fig. 2 shows some Bode plots of equilibrium and their corresponding non-equilibrium impedance measurements. The absolute value of Z at the frequency where the phase angle is zero represents the total ohmic resistance of the cell mainly determined by the resistance of the electrolyte. Following Ohm's law the electrolyte resistance decreases from about 1.6 mΩ at 10 °C to 1.2 mΩ at 50 °C. The structure of the inductive part of the spectra in the high-frequency range ($\varphi > 0°$) is basically independent of the temperature. The mid-frequency ranges from about 1 Hz to 1 kHz represents the charge transfer kinetics at the electrode/electrolyte interface and shows a strong temperature dependence, which can be expressed by the Butler–Volmer equation.

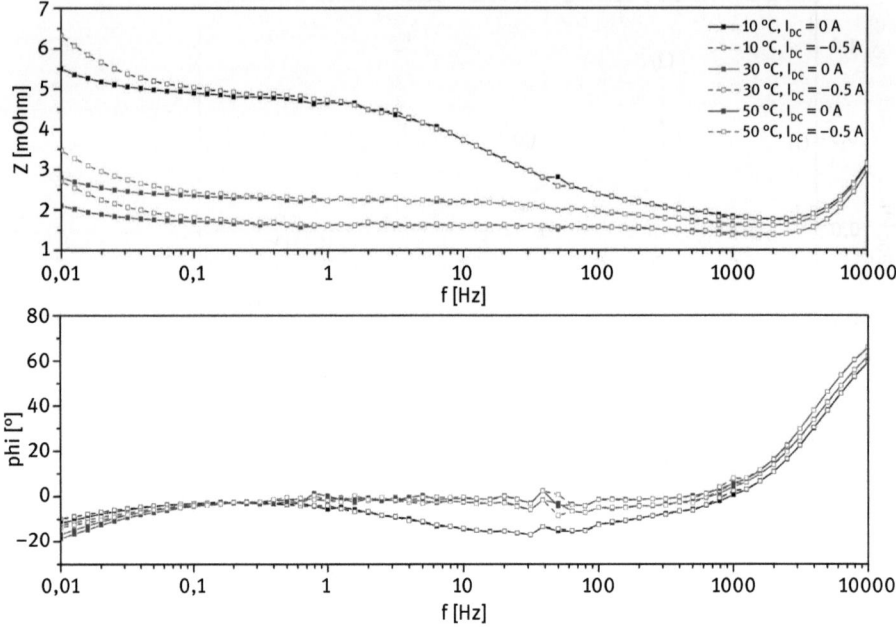

Fig. 2. Bode plot of impedance measurements with $I_{DC} = 0$ A and $I_{DC} = -0.5$ A at 10, 30 and 50 °C, respectively.

Below 1 Hz diffusion is the dominating process. The qualitative behaviour of the impedance is almost equal in the equilibrium and non-equilibrium spectra, respectively. This indicates that the measurements were performed under reproducible conditions. The qualitative behaviour of the impedance as a function of the frequency and the temperature is comparable to the results of other impedance studies on lithium-ion cells [2–6]. Applying a DC-offset during the impedance measurements leads to changing impedances indicating that the direct current influences the diffusion kinetics.

3.2 In-situ Impedance Measurements

The precise quantification of the correlation between the battery temperature and its impedance is the most important requirement for an accurate temperature estimation. Besides a quantification of this correlation for electrochemical and thermal equilibrium state, a massive amount of characterization effort as well as increased model complexity would arise if the correlation has to be quantified separately for all non-equilibrium states. Thus, the excitation frequencies at which the correlation is quantified should be carefully chosen in a way that the influence of the non-equilibrium states is minimized. This would only lead to the necessity of the numerical correlation gathered during equilibrium state. Using this correlation in non-equilibrium conditions leads to an error of the temperature estimation depending on the deviation of the battery impedance between equilibrium and non-equilibrium states, which will be discussed below.

Fig. 3A shows the temperature dependence of the impedance Z measured at a chosen frequency under the different load conditions described in Section 3.1. The relations between the impedance and the temperature can be accurately fitted by polynomial functions, whereas $Z_{eq}(T)$ is the correlation function obtained in equilibrium state and $Z_{neq}(T, I_{DC}, ...)$ are the correlation functions for different non-equilibrium conditions. The deviations between the impedances obtained at equilibrium and non-equilibrium states are shown in Fig. 3B. It can be seen that the current load, which was superimposed to the impedance measurement, leads to a deviation of less than 4% in a temperature range from 10 °C to 50 °C and about 7% at 0 °C. Applying a current pulse right before an impedance measurement with $I_{DC} = 0$ A leads to a deviation of less than 4% with respect to the equilibrium impedance in the full temperature range.

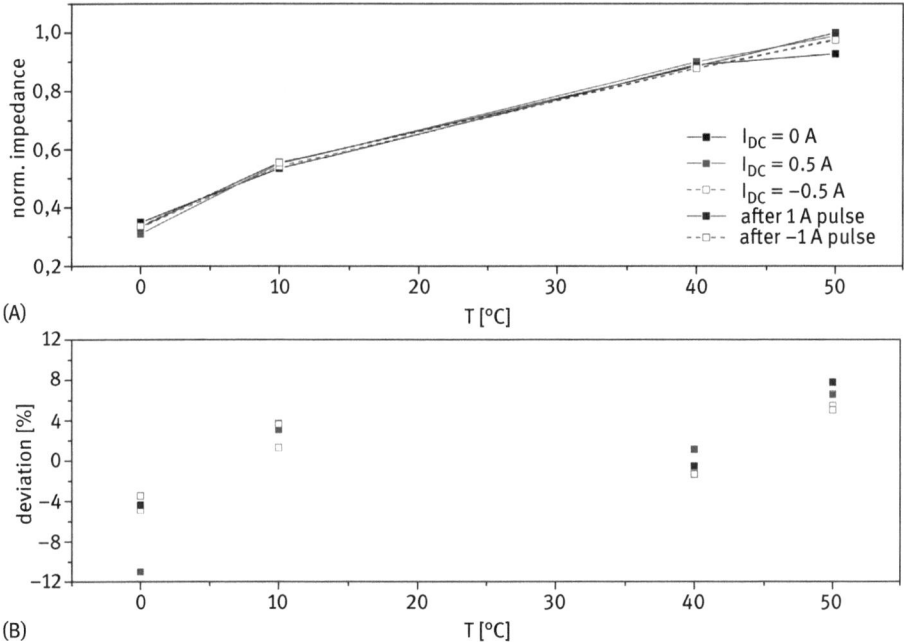

Fig. 3. (A) Measured impedance as a function of battery temperature for different current loads; (B) deviation of the measured impedances with respect to the reference measurement with $I_{DC} = 0$ A at different temperatures.

An estimation of the battery temperature via impedance measurements can be performed using the inverse correlation function $T_{eq}(Z)$ even for impedance values obtained under non-equilibrium conditions. As a consequence thereof the error ΔT of this estimation is given by the difference between $T_{eq}(Z(I_{DC} \neq 0))$ and $T_{eq}(Z(I_{DC} = 0))$ for the measured value of the impedance. Because of the small deviation between $Z_{eq}(T)$ and $Z_{neq}(T, I_{DC}, ...)$ as shown in Fig. 3, this procedure seems suitable for an accurate in-situ temperature estimation.

4 Conclusions

In this chapter an impedance measurement study on automotive-grade Li(NMC)O$_2$ batteries was presented. Impedance measurements in equilibrium and non-equilibrium states were performed at different temperatures. It could be shown that a numerical correlation between the impedance and the battery temperature exists for the examined cell type. The impedance values at chosen frequencies measured at non-equilibrium states showed only a small deviation from those measured at equilibrium state. However, the behaviour of the impedance at higher DC-offsets has to be further investigated to fit application-related circumstances. Thus, temperature estimation with impedance values measured at non-equilibrium conditions can be made using the numerical correlation between the impedance measured in equilibrium state and the battery temperature. This correlation can be generated experimentally in a short period of time saving lots of experimental efforts. This procedure allows to determine the battery temperature by an impedance measurement during dynamic charge and discharge processes, which occurs for example in electric cars. The presented approach can be used in such applications as a redundant method for battery temperature estimation, which has the potential to increase the functional safety of the whole battery pack.

Acknowledgements: The authors acknowledge the financial support by the European Union within the project FP7-2013-ICT-GC/608756.

5 References

[1] W. Waaag, C. Fleischer, and D. U. Sauer, "Critical review of the methods for monitoring of lithium-ion batteries in electric and hybrid vehicles," Journal of Power Sources, vol. 258, pp. 321–339, 2014.

[2] N. S. Spinner, C. T. Love, S. L. Rose-Pehrsson, et al., "Expanding the operational limits of the single-point impedance diagnostic for internal temperature monitoring of lithium-ion batteries," Electrochimica Acta, vol. 174, pp. 488–493, 2015.

[3] R. R. Richardson, P. T. Ireland, and D. A. Howey, "Battery internal temperature estimation by combined impedance and surface temperature measurement," Journal of Power Sources, vol. 265, pp. 254–261, 2014.

[4] J. P. Schmidt, S. Arnold, A. Loges, et al., "Measurement of the internal cell temperature via impedance: evaluation and application of a new method," Journal of Power Sources, vol. 243, pp. 110–117, 2013.

[5] J. G. Zhu, Z. C. Sun, X. Z. Wei, et al., "A new lithium-ion battery internal temperature on-line estimate method based on electrochemical impedance spectroscopy measurement," Journal of Power Sources, vol. 271, pp. 990–1004, 2015.

[6] L. H. J. Raijmakers, D. L. Danilov, J. P. M. van Lammeren, et al., "Sensorless battery temperature measurements based on electrochemical impedance spectroscopy," Journal of Power Sources, vol. 247, pp. 539–544, 2014.

[7] A. Eddahech, O. Briat, E. Woirgard, et al., "Remaining useful life prediction of lithium batteries in calendar ageing for automotive applications," Microelectronics Reliability, vol. 52, pp. 2438–2442, 2012.

[8] A. Barre, B. Deguilhem, S. Grolleau, et al., "A review on lithium-ion battery ageing mechanisms and estimations for automotive applications," Journal of Power Sources, vol. 241, pp. 680–689, 2013.

[9] J. Xu, C. C. Mi, B. Cao, et al., "A new method to estimate the state of charge of lithium-ion batteries based on the battery impedance model," Journal of Power Sources, vol. 233, pp. 277–284, 2013.

Part II: **Sensors**

Christian Weber, Markus Tahedl and Olfa Kanoun

Capacitive Measurements for Characterizing Thin Layers of Aqueous Solutions

Abstract: In many industrial applications, non-contacting capacitive sensors are used to detect conductive fluids. In some applications, online determination of material properties is needed to ensure safe and reliable operation of the sensor. Impedance spectroscopy is an interesting method to characterize the electrode impedance over frequency, conductivity and layer thickness of the fluid. In this contribution, a measurement set-up to generate thin layers of aqueous solutions is described and evaluated. Impedance spectra for different conductivities and layer thicknesses were recorded. The measured data were checked for consistency using a Kramers–Kronig fitting algorithm. Data are consistent in a frequency range from 20 kHz up to 30 MHz. Measured data are interpreted using an electrical equivalent circuit. The proposed circuit fits well to experimental data.

Keywords: Impedance sensors, capacitive sensors, electrical equivalent circuit, modeling, consistency check

1 Introduction

Capacitive sensors are often used for contact-less detection of liquid, conductive media. In some applications, the sensors are mounted on the outside of a non-conductive container to detect whether the fluid has exceeded a certain fill level threshold. Conductive films on the inside of the container might influence the sensors operation and produce a false-positive signal. To distinguish a thin conductive film from the actual fill level, one possible approach is to extract material thickness and conductivity from the measured data.

Many approaches to determine material properties from measured impedances have been proposed. An overview of techniques for contact-less measurement of mostly non-conductive fluids is given in [1]. These methods include electronic capacitance tomography, which is used to determine the permittivity distribution from measured data obtained from a system of electrodes for gas/water/oil flow measurement [2]. Other methods include the determination of fill levels using contacting electrode arrays [3] or non-contacting electrode pairs [4]. In this contribution, a measurement set-up for producing thin layers of aqueous solutions is developed. To identify systematic deviations and inconsistencies in the measured data, data validation using Kramers–Kronig relations is carried out. Impedance measurements for different conductivities and layer thicknesses are made and results are evaluated

---

Christian Weber and Markus Tahedl, ifm efector gmbh, 88069 Tettnang, Germany
Olfa Kanoun, Technische Universität Chemnitz, 09126 Chemnitz, Germany
DOI 10.1515/9783110449822-007

using an electrical equivalent circuit (EEC) representing the physical properties of the measurement set-up.

2 Measurement Set-up

Capacitive sensors are often used to detect conductive fluids in large containers. These sensors are compactly built using only a single sensing electrode and a shielding electrode. Capacitance is measured between the sensing electrode and its environment. These sensors may be influenced by thin conductive layers in the vicinity of the sensing electrode. The measurement set-up needs to emulate these conditions as closely as possible.

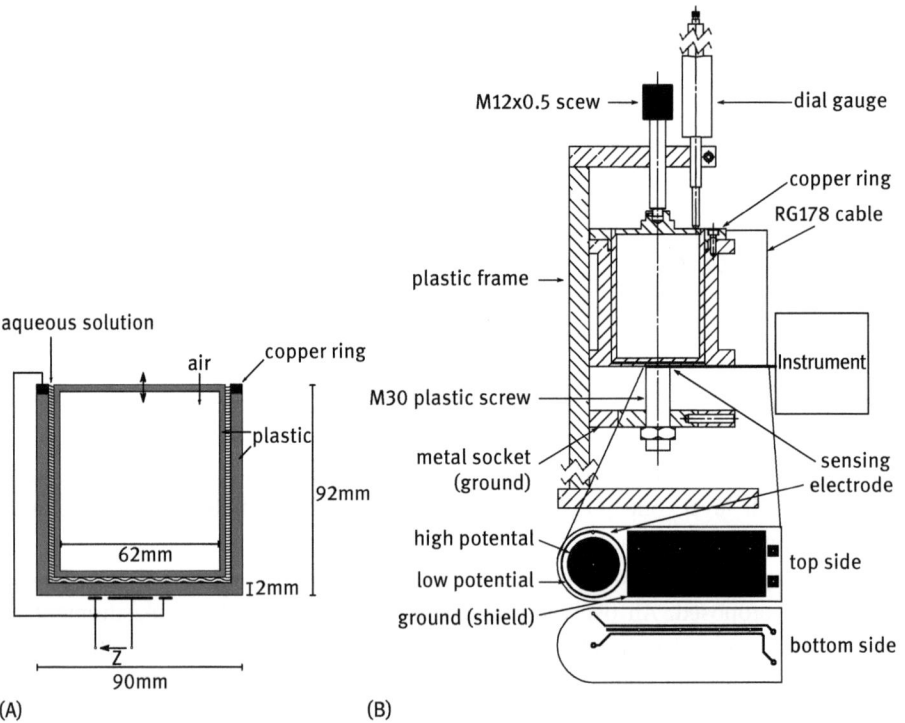

Fig. 1. Measurement set-up concept drawings. (A) The basic concept of the set-up consisting of an inner and an outer cylinder. (B) Final set-up including the sensing electrode on a printed circuit board connecting it to the instrument. The ground potential of the instrument is used as shield potential.

To characterize aqueous films in terms of layer thickness and conductivity, the measurement set-up needs to reliably produce a film of the solution with a defined thickness. Influence of the measurement set-up needs to be negligible. Also, the set-up must be easily connectable to an impedance measurement instrument.

Fig. 2. Photos of the measurement set-up. (A) An overview of the set-up. (B) PCB electrode connected to the instrument via short wires using 16047E adapter.

The measurement set-up shown in Fig. 1 consists of an outer and an inner cylinder fabricated from polyoxymethylene (POM) and polyethylenterephthalate (PET), respectively. The measurement set-up is rotationally symmetric, because it is easier to model with either an analytical model [5] or the finite element method. Fig. 2 shows two photographs of the measurement set-up. The diameter of the outer cylinder was chosen to be much larger than the electrode diameter to emulate a large container. The inner cylinder is hollow to emulate the free space behind the liquid layer. The outer cylinder contains the solution to be measured and a copper ring, used for grounding the solution, which is connected to the low-port of an Agilent 4294A Impedance Analyser via a RG178 coaxial cable. To adjust the layer thickness at the bottom of the set-up, the inner cylinder is connected to a M12 × 0.5 screw, while its vertical position is measured with a dial gauge.

The sensing electrode is circular consisting of an inner and an outer electrode with a radius of 9.5 and 11 mm, respectively. The high potential port is connected to the inner sensing electrode, while the low port is connected to the copper ring and the outer sensing electrode. The metal socket is connected to the ground terminal of the instrument to shield the backside of the electrode. Connection to the instrument is established by two printed circuit board (PCB) tracks. A third track between the low- and high-potential track is connected to ground to shield their stray capacitance. A large shielding electrode on the topside of the PCB prevents additional stray effects caused by the tracks.

Tap water and potassium chloride solutions of varying concentration are used as material under test (MUT). Potassium chloride is used because its conductivity is well defined [6]. The measurement set-up containing the MUT is connected to the instrument via an Agilent 16047E adapter. In this chapter, the measurement set-up containing the liquid will be referred to as device under test (DUT).

To eliminate stray parameters of the adapter, an OPEN/SHORT/LOAD calibration scheme provided by the instrument has been employed. A 100Ω resistor was used as LOAD-standard. Its series resistance was measured at 40 Hz, while its series inductance was measured at 2 MHz. Both values were used as parameters for the instrument's built-in calibration procedure. Calibration was done at the adapter without the connected measurement set-up, because stray effects of the PCB tracks are sufficiently small. The stray capacitance and loss factor of the PCB tracks were measured by fabricating a variant of the PCB without an electrode. At 2 MHz the stray capacitance between tracks is 0.27 pF with a loss factor of 8‰. The series resistance is measured by shorting the two PCB tracks at their end. Their resistance at 40 Hz is 0.12Ω and their inductance at 2 MHz is 0.11 μH. The cable capacitance of the RG178 coaxial cable connecting the copper ring to the instruments low port is compensated by connecting the outer conductor to instrument ground. Its stray inductance L_s may be calculated by using the known formula for the impedance of a lossless transmission line Z_L and the stray capacitance C_s:

$$L_s = Z_L^2 \cdot C_s \tag{1}$$

Values for C_s and Z_L were taken from [7]. The stray inductance of an RG178 cable is therefore 0.24 $\frac{\mu H}{m}$.

3 Data Validation

When interpreting data using EECs, two different causes for systematic deviations between measured and fitted data can be distinguished. Either the chosen equivalent circuit is not suitable for a particular data set, or the data set itself contains systematic errors. To check the data set for systematic errors, data validation may be employed. The real and imaginary part of impedance data are related by the Kramers–Kronig relations. If a system fulfills these relations, it is causal, linear and time-invariant. To check the consistency of the measurement, data are validated using a linear Kramers–Kronig fitting algorithm proposed in [8] and improved in [9]. As described in [8], the measured admittance spectrum Y_{meas} is fitted to the circuit in Fig. 3. It consists of M RC-elements with logarithmically distributed time constants. Typical values range from three to seven time constants per decade [9]. The real part of the

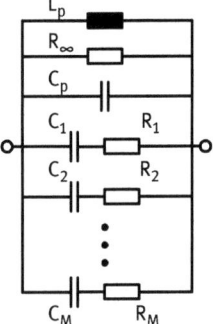

Fig. 3. EEC used for data validation.

model admittance Y_{mod} depends on the angular frequency ω and can be expressed as

$$\Re(Y_{mod}) = \frac{1}{R_\infty} + \sum_{i=1}^{M} \frac{\omega^2 C_i^2 R_i}{1+\omega^2 C_i^2 R_i^2} \quad (2)$$

while the imaginary part is

$$\Im(Y_{mod}) = \omega C_p - \frac{1}{\omega L_p} + \sum_{i=1}^{M} \frac{\omega C_i}{1+\omega^2 C_i^2 R_i^2} \quad (3)$$

The EEC is linear, causal and time-invariant if its parameters are constant. Also, its poles have to be at fixed frequencies and have to be negative. The parameters L_p, C_p, R_∞, C_i and R_i may be extracted by a linear fit. However, the EEC shown in Fig. 3 is a purely behavioural model and does not represent physical processes. The fitted values might even be negative [8]. Therefore, this model is only used for data validation. If the relative difference between measured and fitted data is small, the data set fulfils Kramers–Kronig transformation rules and is therefore valid. The relative difference between fitted and measured data may be computed as follows:

$$\Delta_{Re} = \frac{|\Re(Y_{meas}) - \Re(Y_{mod})|}{|\Re(Y_{meas})|} \cdot 100\% \quad (4)$$

$$\Delta_{Im} = \frac{|\Im(Y_{meas}) - \Im(Y_{mod})|}{|\Im(Y_{meas})|} \cdot 100\% \quad (5)$$

Fig. 4 shows an example of data validation using Kramers–Kronig relations. All measured data points with a relative difference greater than the error threshold Δ_{Re} or Δ_{Im} may be excluded from the spectrum. The error threshold was determined by experiment. Typical values range from 1% to 4%.

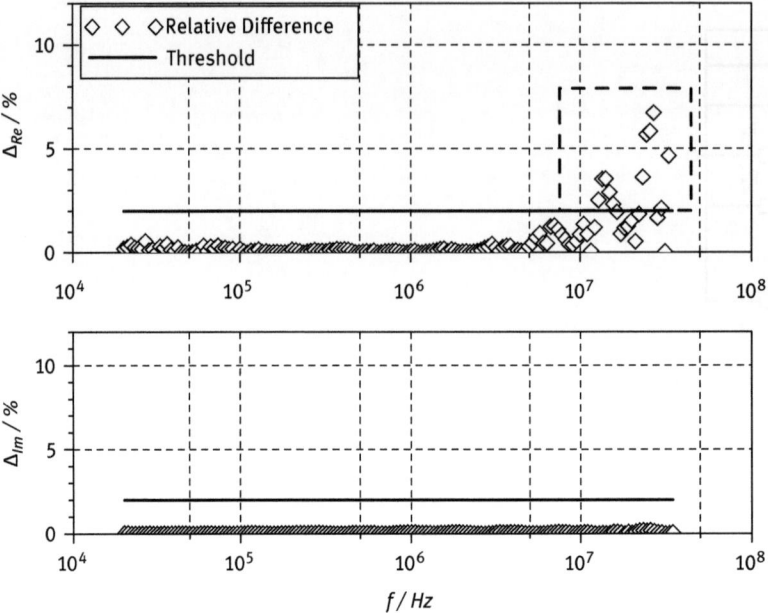

Fig. 4. Example for data validation using Kramers–Kronig relations. The marked data points are excluded for the measured data.

4 Experimental Investigations

Complex Impedance spectra were measured at room temperature for four different layer thicknesses between 0.25 and 1 mm and two conductivities σ in a frequency range from 20 kHz to 30 MHz. At frequencies greater than 30 MHz measured data are no longer consistent. Cause for this effect might be the measurement inaccuracy of the 16047E adapter. It is specified in a frequency range up to 110 MHz but causes a significant measurement error at frequencies greater than 15 MHz [10]. Below 20 kHz the absolute impedance value of the device under test is too large for accurate measurements of $\Re(Z)$. The upper plots of figures 5 and 6 show Nyquist graphs of measured and validated data. At low frequencies, the effect of the lossy dielectric materials used in the measurement set-up and the effect of the PCB is dominant because the electric field cannot penetrate the conductive MUT. Because of this, the coupling capacitance has a much larger impedance than the rest of the components of the DUT. To extract the physical parameters of the system, measured data are fitted to an EEC using [11]. Theoretical considerations [5] suggest a second-order EEC shown in Fig. 7b. The constant phase element (CPE) is used to model lossy dielectric media (e.g. plastic components) used in the measurement set-up. Its complex impedance may be

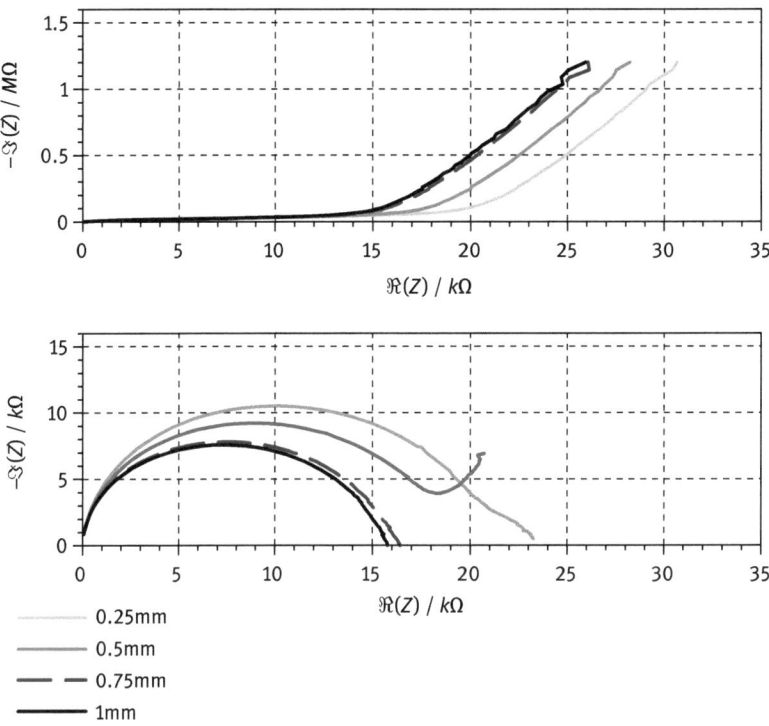

Fig. 5. Nyquist plots for tap water $\sigma = 0.59\,\frac{mS}{cm}$, $T = 21.9°C$. The top plot shows the measured and validated data. The bottom plot shows the same data set minus the constant phase element.

calculated by

$$Z_{CPE} = \frac{1}{Q}(j\omega)^{-n} \qquad (6)$$

with parameters Q, n and $j = \sqrt{-1}$ [12]. The sensor electrode capacitively drives the MUT, which is represented by the contact resistance R_1. The bulk characteristics of the MUT are described by R_2 and C_2. Direct capacitive coupling between the measuring electrode and low potential is modelled by C_1. Subtracting Z_{CPE} from the measured impedance yields the lower plots of Figs 5 and 6. The EEC shown in figure 7b is based on physical considerations. However, when using a non-linear fit, it may be difficult to choose initial values for this circuit. Instead, we suggest to first fit the validated data set to the EEC in Fig. 7A. Initial values for this topology are easier to choose because its time-constants are represented by a series circuit. After extracting the parameters for Fig. 7A, parameters for Fig. 7B may be calculated by considering the transfer function of both EECs. Both circuits have the same number of poles and zeros. If the poles/zeros are equal, both EECs behave the same way, which leads to a system of equations.

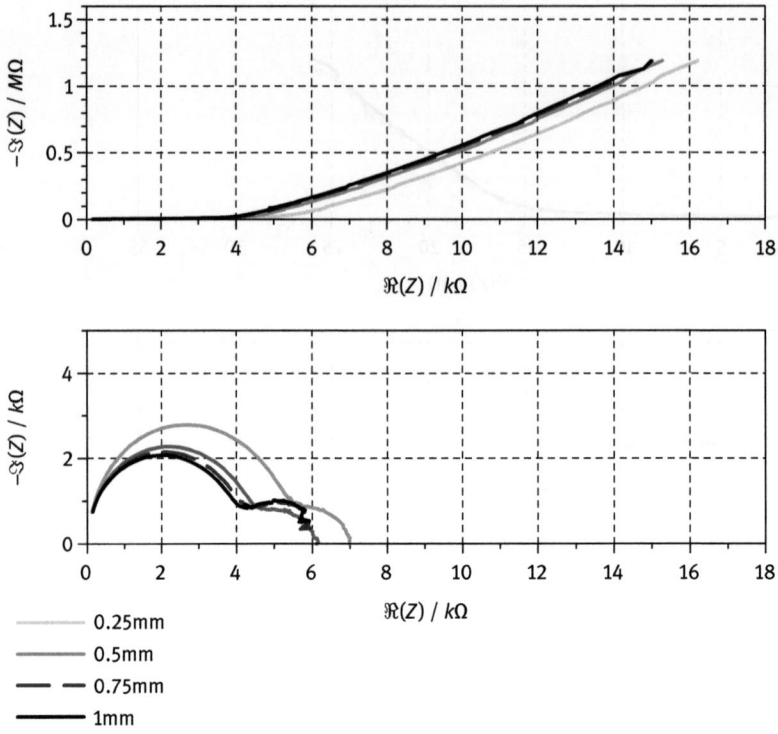

Fig. 6. Nyquist plots for KCl solution $\sigma = 2.17\,\frac{mS}{cm}$, $T = 20.7°C$. The top plot shows the measured and validated data. The bottom plot shows the same data set minus the constant phase element.

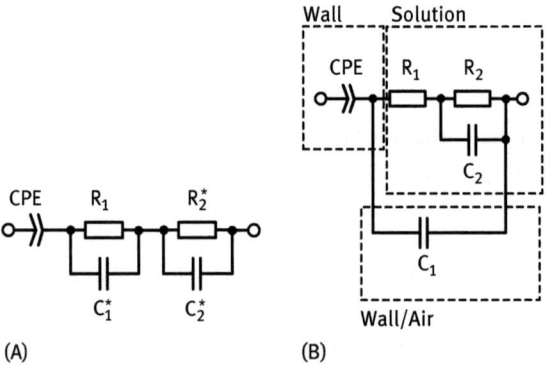

Fig. 7. Equivalent circuit models. The left circuit is used to fit measured data. Fitted parameters are then converted to the right circuit. (A) Behavioural EEC for measured data; (B) EEC representing physical processes.

Solving the equation system yields

$$R_1 = \frac{R_1^* R_2^* \cdot (C_1^* + C_2^*)^2}{R_1^* C_1^{*2} + R_2^* C_2^{*2}} \tag{7}$$

$$R_2 = \frac{(C_1^* R_1^* - C_2^* R_2^*)^2}{R_1^* C_1^{*2} + R_2^* C_2^{*2}} \tag{8}$$

$$C_1 = \frac{C_1^* C_2^*}{C_1^* + C_2^*} \tag{9}$$

$$C_2 = \frac{\left(R_1^* C_1^{*2} + R_2^* C_2^{*2}\right)^2}{(C_1^* + C_2^*) \cdot (C_1^* R_1^* - C_2^* R_2^*)^2} \tag{10}$$

5 Results

The parameters for the EEC are extracted and shown in Tabs. 1 and 2. Figs 8 and 9 show the relative difference Δ of the absolute impedance $|Z|$ and the phase angle φ between fitted curves and the measured data. For all cases, the relative difference is within ± 2%. The simple EEC therefore approximates the general behaviour of measured data

Tab. 1. Fitted parameters for tap water

	Tap Water 0.59 mS/cm			
	0.25 mm	0.5 mm	0.75 mm	1 mm
C_1 / pF	7.3	7.67	7.97	8.03
C_2 / pF	1150	1250	1340	1230
R_1 / kΩ	20.6	17.9	15.6	15.3
R_2 / kΩ	3.04	2.08	1.63	1.91
$Q \cdot 10^{12}$	6.85	6.89	6.92	6.91
n	0.996	0.996	0.995	0.995

Tab. 2. Fitted parameters for potassium chloride

	KCl 2.17 mS/cm			
	0.25 mm	0.5 mm	0.75 mm	1 mm
C_1 / pF	7.58	8.07	8.1	8.2
C_2 / pF	915	884	894	903
R_1 / kΩ	5.53	4.57	4.31	4.16
R_2 / kΩ	1.7	1.73	1.65	1.62
$Q \cdot 10^{12}$	7	7	7.01	7.01
n	0.995	0.995	0.995	0.995

reasonably well. However, there is still a systematic difference between the measured and fitted curves since the relative difference is not a random distribution around the 0% axis. The deviation is highest in $\Re(Z)$, where the relative difference between fitted and measured data is up to 20%. Because the real part of the measured impedance is much smaller than the imaginary part, the general behaviour of the system can still be approximated well.

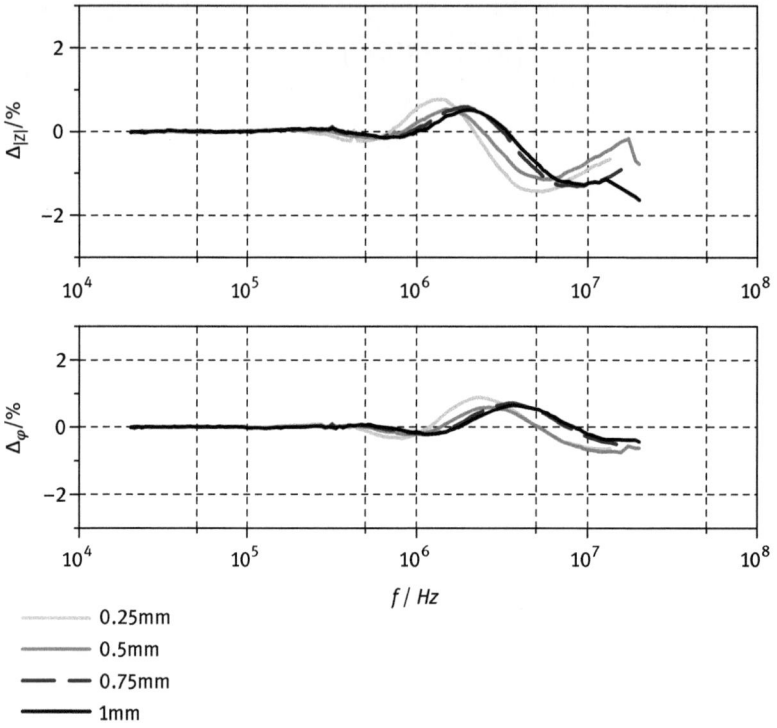

Fig. 8. Relative difference between fitted and measured data (tap water). (Top plot: Relative difference for |Z|. Bottom plot: Relative difference for arg(Z)).

To check whether the chosen EEC yields enough information about the physical properties of the system, one has to closely consider the extracted parameters.

The CPE parameters are approximately constant for all layer thicknesses and both measured conductivities. Also, n is very close to 1, which is to be expected when employing capacitive coupling, although some losses may arise from plastics and the PCB. The slight dependence of the CPE parameters on MUT properties may arise from stray field effects of the measurement electrode through the thin outer wall of the measurement set-up. Also, for constant layer thicknesses, changing material conductivity should only affect the fitted resistances, which is not the case here. One would also expect $(R_1 + R_2) \cdot \sigma$ to be constant for a single-layer thickness [5]. The extracted capacitances do not always increase monotonously with increasing layer thickness. A possible cause for this may be frequency-dependent conductivity and permittivity of the MUT or insufficient sensitivity of the non-linear fit.

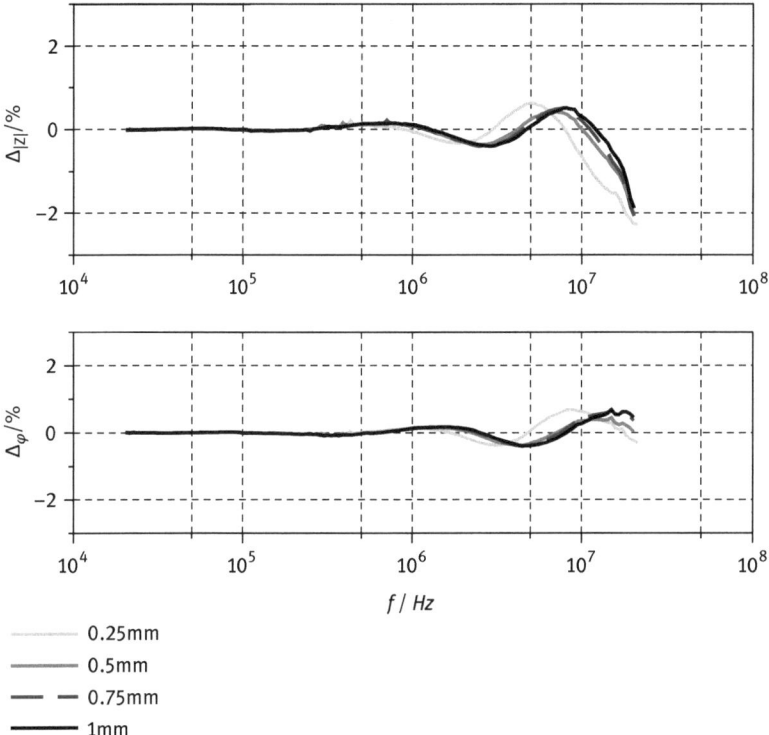

Fig. 9. Relative difference between fitted and measured data (potassium chloride). (Top plot: Relative difference for $|Z|$. Bottom plot: Relative difference for $\arg(Z)$).

To check whether the sensitivity of the non-linear fit is sufficient to extract the desired parameters, a sensitivity analysis [13] for the EEC in Fig. 7A is carried out. The sensitivity of an EEC towards a certain parameter may be calculated by considering the partial derivation of the model function $f(x_1, x_2, ..., x_N)$. The normalized sensitivity S_{x_i} for a parameter x_i is [13]

$$S_{x_i} = \frac{\partial f}{\partial x_i} \cdot x_i \qquad (11)$$

The result of the sensitivity analysis for the given EEC in Fig. 7A is shown in Figs 10 and 11. In the low-frequency region, the sensitivity of the CPE parameters is several orders

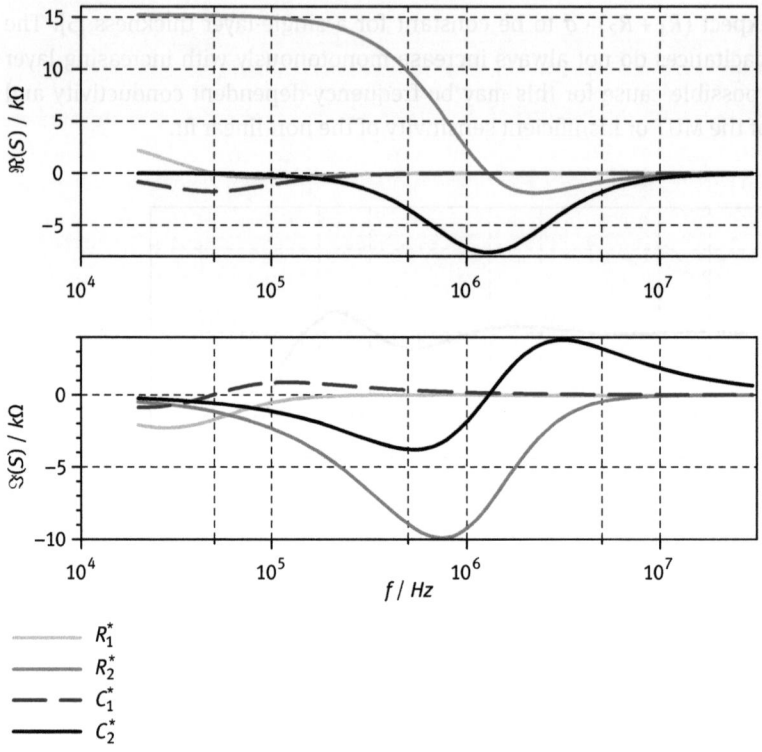

Fig. 10. Sensitivity for R_1^*, R_2^*, C_1^* and C_2^*. The top and bottom plots show the sensitivity for real and imaginary parts, respectively.

of magnitude higher than the sensitivity of all other parameters. Also, the greater of the two time constants made up by R_1^* and C_1^* has a much lower sensitivity, than the time constant made up by R_2^* and C_2^*. Its sensitivity approaches zero for frequencies greater than 500 kHz. Thus, to reliably fit R_1^* and C_1^*, more measurement points in the low-frequency region are needed.

6 Conclusion

A measurement set-up for characterizing the electric impedance of thin layers of aqueous solutions has been developed and measurements have been carried out. Data consistency is observed in a frequency range from 20 kHz up to 30 MHz. Measured data in this frequency range are consistent in terms of linearity, causality and time invariance. This range may be used to characterize aqueous solutions with a conductivity of up to 2.17 mS/cm and layer thickness of up to 1 mm. Plastic components

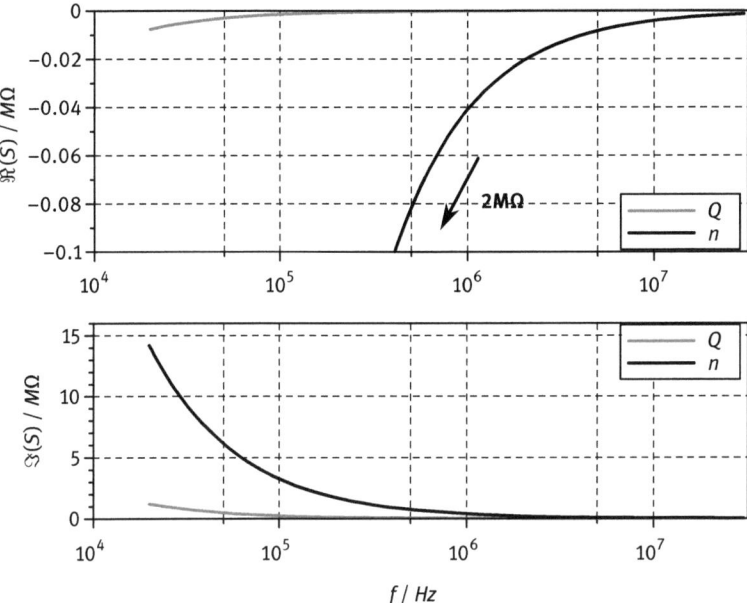

Fig. 11. Sensitivity for CPE parameters Q and n. The top and bottom plots show the sensitivity for real and imaginary part, respectively.

of the set-up can be modelled as a CPE and subtracted from the measured spectrum. The data may be fitted to an EEC with two time constants modelling the conductivity and layer thickness of the MUT in terms of electrical resistance and capacitance. The model approximates the behaviour of the system reasonably well. However, there is still a systematic deviation between modelled and fitted data. Causes for this effect will be investigated. At frequencies greater than 30 MHz, measured data are no longer consistent because the relative difference between measured and Kramers–Kronig fitted data exceeds the error threshold. Further investigation regarding measurement artefacts will be performed.

7 References

[1] H. Zangl, A. Fuchs, and T. Bretterklieber, "Non-invasive measurements of fluids by means of capacitive sensors," Elektrotechnik und Informationstechnik, vol. 126, no. 1–2, pp. 8–12, 2009.

[2] Y. Li, W. Yang, Z. Wu, et al., "Gas/oil/water flow measurement by electrical capacitance tomography," in 2012 IEEE International Conference on Imaging Systems and Techniques (IST), July 2012, Manchester, UK, pp. 83–88.

[3] R. C. Alonso, "Liquid interface level sensing using electrode arrays," PhD thesis, UPC BarcelonaTech, 2007.

[4] K. Biswas, S. Sen, and P. K. Dutta, "Modeling of a capacitive probe in a polarizable medium," Sensors and Actuators A: Physical, vol. 120, no. 1, pp. 115–122, 2005.

[5] C. Weber, F. Wendler, M. Tahedl, et al., "2D-Modellierung des Einflusses leitfähiger Schichten auf die Sensorimpedanz bei kapazitiven Sensoren," tm - Technisches Messen, vol. 81, no. 9, pp. 442–449, 2014.

[6] R. H. Shreiner and K. W. Pratt, Primary Standards and Standard Reference Materials for Electrolytic Conductivity. U.S. Department of Commerce, Technology Administration, National Institute of Standards and Technology, 2004.

[7] RG178B/U Technical Data Sheet, Draka, [Online], March 2015. Available at: http://www.ok1rr.com/coax/50/RG178B-U.pdf.

[8] B. A. Boukamp, "A linear Kronig-Kramers transform for immitance data validation," Journal of the Electrochemical Society, vol. 142, no. 6, pp. 1885–1894, 1995.

[9] U. Tröltzsch, "Modellbasierte Zustandsdiagnose von Gerätebatterien," PhD thesis, Universität der Bundeswehr München, 2005.

[10] 16047E Test Fixture, 40 Hz to 110 MHz Operation and Service Manual, Keysight Technologies, [Online], March 2015. Available at: http://www.keysight.com/main/editorial.jspx?cc=DE&lc=ger&ckey=1000002127:epsg:man&nid=-34051.536880747.00&id=1000002127:epsg:man.

[11] A. S. Bondarenko and G. Ragoisha, Potentiodynamic Electrochemical Impedance Spectroscopy, A. L. Pomerantsev, Ed. New York, NY: Nova Science Publishers, 2005, pp. 89–102.

[12] J. R. Macdonald, "Note on the parameterization of the constant-phase admittance element," Solid State Ionics, vol. 13, pp. 147–149, 1984.

[13] O. Kanoun, U. Tröltzsch, and H.-R. Tränkler, "Benefits of evolutionary strategy in modeling of impedance spectra," Electrochimica Acta, vol. 51, no. 8–9, pp. 1453–1461, 2006.

Łukasz Macioszek and Ryszard Rybski

Low-Frequency Dielectric Spectroscopy Approach to Water Content in Winter Premium Diesel Fuel Assessment

Abstract: In this chapter, an attempt was made to evaluate the water content in the diesel fuel using dielectric constant calculated on the basis of diesel's impedance. Measurements were performed with the use of electrochemical impedance spectroscopy system in frequency range of 0.1 Hz to 1 kHz and were designed to verify experimentally that this approach can be used as an in situ, relatively cheap and fast alternative compared with laboratory method widely used nowadays. Known amounts of water were added to the diesel samples beyond content allowed by the standard. Experimental results show that with some limitations, impedance spectroscopy and dielectric constant calculation can be used to estimate water content in diesel fuel at least until a solubility limit is reached.

Keywords: Dielectric spectroscopy, diesel fuel, water content

1 Introduction

Many types of vehicles, locomotives, ships and even electricity generation plants are driven by a very common fuel. Diesel is a result of crude oil's distillation, and it consists of several types of hydrocarbons (about 75% saturated, 25% aromatic), as well as various kinds of additives and enhancers added during production [1]. A few percent of fatty acid methyl ester are also added in many countries. For example, in Poland, there are 7% (V/V) of biocomponents in each type of commercially available fuel.

The precise chemical composition of diesel is never the same. One of the reasons is that distillation, like any process, is never performed with 100% efficiency. Another one is the crude oil itself, which is not always the same. At the opposite, there is an obvious need to maintain the quality of diesel fuel that is being sold. Within the European Union, the EN 590 standard is valid. It contains ranges of physical properties' permissible values and lists the corresponding test methods to verify them.

Improved performance and efficiency of diesel engines imply an increase in their sensitivity to all kinds of the contaminants in fuel. Very high pressures exceeding 300 MPa are featured by modern direct injection systems (common rail). Such values

---

Łukasz Macioszek and Ryszard Rybski, Institute of Metrology, Electronics and Computer Science, Faculty of Computer, Electrical and Control Engineering University of Zielona Góra, Zielona Góra, Poland

would not be possible to achieve without rigorous fuel parameters that ensure the absence of aggressive for metal alloys substances, of which engines are composed. Particularly newer engines' fuel rails are sensitive to water content in the fuel. Although serious failures due to water contamination may occur relatively rarely, they do happen creating considerable and unnecessary expenses.

Water and diesel are immiscible in general, but over a dozen ppm at a production stage is absolutely normal and difficult to decrease. From a consumer point of view, higher water amounts can be present in fuel because of either condensation processes or some kind of failure during fuel transportation or storage. Denser water settles at the bottom over time, but each movement can pick up some amount again. The solubility limit of water in diesel is reported to be very low (100 ppm at 40°C [2]), but various kinds of additives can make it significantly higher. Although such amount is relatively low, water in diesel can still be treated as a dangerous contamination. Only 200 mg/kg is allowed by the aforementioned standard, EN 590. Coulometric Karl Fischer titration method is indicated to be used for this determination, as it allows obtaining very precise results in the range of 0.003%–0.100% (m/m). However, it cannot be used in situ as it requires preparation of test samples and needs to be performed in laboratory. One of the various methods proposed to evaluation of diesel fuel properties is dielectric spectroscopy [3], but it has not yet been used for water content assessment.

Approach presented in this chapter focuses on being relatively easy to be implemented on-site in the future. For this a lower frequency range is chosen as it may be both sufficient in terms of obtained results and less demanding in hardware and immunity to various types of noise when used in portable measuring system. Reported measurements should be regarded as preliminary, which were aimed to identify areas requiring more attention in order to improve received results. Treating diesel fuel as dielectric may be similar to what is done to transformer oil. In that case, using broadband dielectric spectroscopy is not necessary [4]. Determination of water content in diesel fuel by calculating its dielectric constant may be another method that may be complementary or replaceable for fitting equivalent circuit elements' values [5]. In both cases, impedance spectroscopy may be used to study properties of diesel fuel.

2 Material and Methods

The main objective of the study was to verify experimentally that different known water contents in the diesel fuel are identifiable and thus measurable using impedance spectroscopy and calculated dielectric constant. Winter premium diesel fuel samples were obtained from a gas station and were examined before and after water addition. Bought sample was sent for an analysis by the titration method, and the result shows that there was 40 mg/kg of water content from the moment of fuel's purchase [6].

2.1 Diesel Fuel Samples

Premium-type diesel fuels are declared by producers as containing engine cleaning substances, with a higher cetane number or as sulphur free. Some are advertised as being capable of protecting engines against corrosion or improving their lubricity. Diesel studies are difficult because of the lack of full reproducibility of fuel itself combined with the fact that each additive and enhancer added during or after production are secrets of the producer. Pure diesel fuel can be treated as dielectric [7], but each additive makes its dielectric properties more complex.

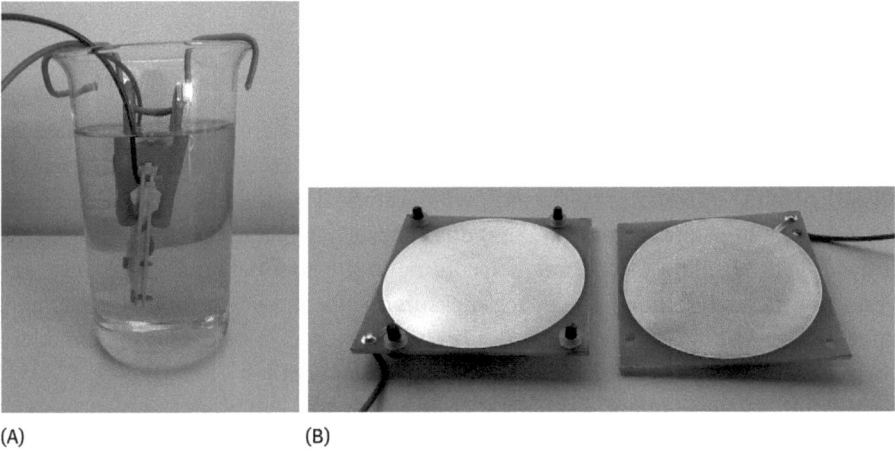

(A)   (B)

Fig. 1. (A) Sample with immersed electrodes. (B) Electrodes used in the experiment.

Samples of 500 ml capacity were examined in a glass beakers (shown in Fig. 1A) at a temperature of 27.5 °C. Demineralized water was added to bought fuel to create separate samples with concentrations of 0.01%, 0.02%, 0.03%, 0.04%, 0.05% and 0.06% (in volume). All samples were stirred with a magnetic stirrer for 30 minutes before starting and during the experiment at the same velocity to ensure the same measurement conditions. More efficient methods of mixing immiscible substances exist, but this mechanic way was supposed to mimic what can be observed in everyday life, for example, in car fuel tanks.

2.2 Electrodes

Two circular parallel electrodes made of copper with a surface area of 25 cm^2 each, spaced by 1.95 mm, were immersed in diesel samples. Fig. 1B illustrates both electrodes and four supports providing the planned distance. As it can be seen, the actual electrodes were in form of copper layer (35 μm) on a laminate, the same that is usually used to build electronic circuits. Although better materials that can be used as electrodes exist, such as platinum, gold or stainless steel, which are less chemically reactive, the ones presented in the chapter were good enough for performing preliminary measurements. After each measurement, electrodes were washed with water and acetone to clear out all the remains of diesel oil. To maintain reproducibility of the measurements, electrodes were also polished with water emery paper to remove any oxides. This process changes micro-scale properties of copper but in macro-scale had no negative influence on measured impedance. After such cleaning, electrodes were dried and used again.

2.3 Impedance Spectroscopy

Prepared diesel fuel samples' impedance was measured in a frequency range of 0.1 Hz to 1 kHz with a 300-mV RMS amplitude. The EG&G/Princeton Applied Research electrochemical impedance spectroscopy laboratory system was used, consisting of the 263A potentiostat/galvanostat, 5210 dual-phase lock-in amplifier and PowerSINE software. The system was calibrated according to the manufacturer's recommendations, and self-calibration before each measurement was also performed. Measured impedance in form of $Z^*(\omega) = Z'(\omega) - jZ''(\omega)$ was used to calculate the samples' complex relative permittivity with the use of electric modulus [7] $M^*(\omega) = 1/\varepsilon^*(\omega) = j\omega C_0 Z^*(\omega)$, where C_0 refers to capacitance of the vacuum cell.

3 Results and Discussion

The aforementioned percentage values of water added to diesel fuel samples up to 0.06% conform to 40 (bought fuel, no added water), 160, 281, 401, 522, 642 and 763 mg/kg, respectively. It means that the amount of water permitted by the EN 590 standard is exceeded in third and subsequent samples. Complex phenomena happen when water is added to diesel fuel, as it is not homogeneous substance. More reactive to water than fuel itself can be additives added during production to increase its conductivity to above 50 pS/m needed by transportation standard. Additives used are the secret of producer and can be of various types; thus, it is hard or impossible to make assumptions of precise electrochemical reactions taking place in water-contaminated diesel. Diesel's multifarious composition was the reason to expect stronger increase in dielectric constant with the increase of water amount than just result of calculation by adding two different values (34 ÷ 88 for water and about 2 for pure diesel) in the above-mentioned proportions.

Stray effects can be omitted as several measurement repeats did not reveal values exceeding measurement error of the system used, and the most obvious factor affecting value, that is temperature, was maintained nearly constant (±0.1 °C). Fig. 2 presents the Nyquist plot with the measured impedance of seven diesel samples. It can be seen that measured values form single semicircles. The centres of semicircles are a bit depressed, which suggests that calculated capacitance will vary with frequency. Although the right ends of semicircles are a bit raised, a further decrease in frequency did not reveal second semicircles (not shown in Fig. 2). This raise may be associated with diffusion phenomena occurring in the dielectric when a copper electrode is being used [8] and was not treated as a primary and dominant phenomenon.

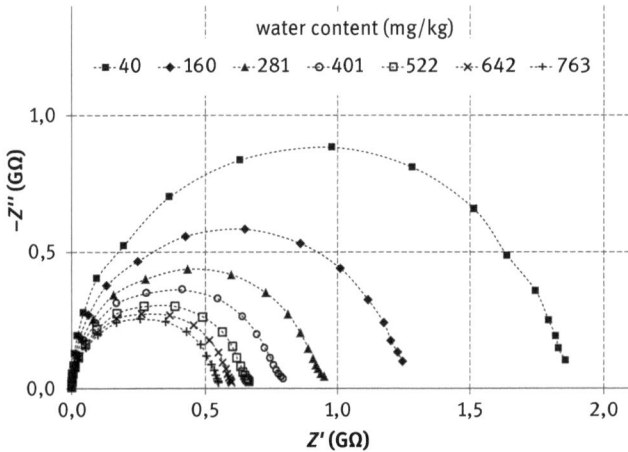

Fig. 2. Nyquist plot of examined diesel fuel samples containing listed water concentrations.

Note that there was no apparent difference between diesel fuel samples containing 763 mg/kg and higher amounts of water, which are not presented in this chapter. It can be seen in Fig. 3, where the real part of calculated dielectric constant (ε') is displayed that the biggest difference between values happens to samples with 40 and 160 mg/kg of water. Further water addition made the differences between samples less visible but until then dielectric constant of solution versus water content seems to be close to linear. Fig. 3 shows also that calculated dielectric constant of above 6 might be the mark of dangerously contaminated fuel. Water content of about 800 mg/kg may be water solubility limit in this very examined winter premium diesel. Higher concentrations (i.e., >2 g/kg) resulted in a visible change in transparency. Complex electrochemical phenomena occur in non-homogeneous, immiscible liquids that are studied as an interface between two immiscible electrolyte solutions (ITIES) and were not within the scope of this chapter. Further increase in frequency up to 100 kHz did not reveal visible differences between samples' measured impedance values. Thus, it may be stated that in case of water-contaminated diesel, it may be pointless to increase frequency above kilohertz range, which is also seen in studying dielectric measurements of transformer oil [4].

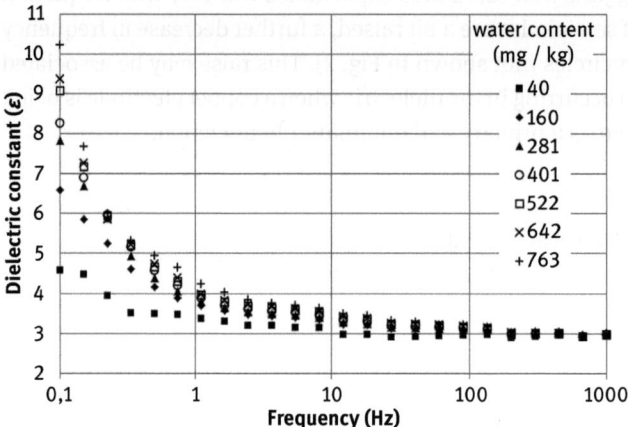

Fig. 3. Frequency spectra of diesel samples' dielectric constant (ε').

Fig. 4 illustrates the calibration curve with a limit of detection at about 800 mg/kg of water content in diesel fuel. Although till about 1.5 g/kg of added water samples was at first sight homogeneous, their measured impedance values and thus dielectric constants were indistinguishable.

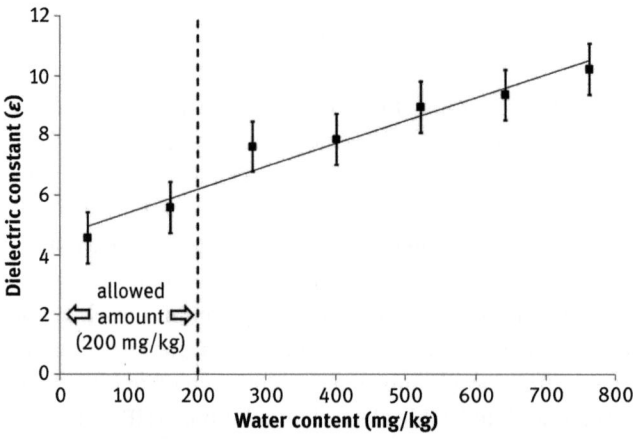

Fig. 4. Water content dependence of the dielectric constant (ε') from the studied diesel samples.

4 Conclusion

In this chapter, the use of dielectric spectroscopy for water content in winter premium diesel fuel assessment was discussed. Experimental results were reported and analysed. Calculated dielectric constant of diesel samples as well as approximated calibration curve was presented. They together show that with some limitations, dielectric spectroscopy can be used for the evaluation of water amount in diesel fuel.

The first limitation was assumption within the evaluation that only water addition can change the impedance and thus dielectric constant of diesel fuel in the above-presented way. As the exact composition of each fuel commercially available is a trade secret, it is impossible to define what kind of additional substances and in what quantities were added as improvers during production and what is their effect in electrochemical and dielectric studies of diesel. There are substances that can significantly increase the solubility limit of water in diesel and other that can decrease its resistivity. Examining fuel with such addition would answer the question whether dissolved water can still be easily detectable and measurable with the use of impedance and dielectric spectroscopy.

The second limitation is the cell geometry. Precise capacitance measurements require the smallest possible gap between plate electrodes. Used distance of 1.95 mm may be too large to maintain high measure accuracy but was initially chosen to avoid situation when water droplets would stick to both electrodes and make a short circuit. The diameter of water drops depends on the quantity of water in diesel. It is possible that a 1-mm gap would be sufficient to measure water content up to about 800 mg/kg. In addition, the design of electrodes including Kelvin guard ring could make capacitance measurements even more precise [9].

Another encountered limitation was electrodes made of oxidizing metal. Although the measuring voltage was relatively low (300 mV root mean square), it might be sufficient to speed up the oxidation of electrodes if higher water contents were to be measured. Oxidation effect can cause the lack of measurement repeatability; thus, the proper electrode materials should be used. Evaluation of the best possible material was not within the scope of this chapter but is inevitable. Circular electrodes made of stainless steel were used elsewhere [3] without reports about their oxidation. However, they were not exposed to water. Still, it would be worth checking their usability in water content assessment as well as other kinds of electrodes with gold and platinum that are used widely in electrochemistry.

Given the above limitations, we can still state that dielectric spectroscopy of water-contaminated diesel fuel is reasonable and comparable favourably with methods that require preparation of the samples and thus can only be performed in laboratory. Achieved accuracy of water content assessment may not reach the same high level but it is not the most important factor in each case. Even though the top measurable content of water was not established, contents exceeding the amount permitted by EN 590 standard were examined. Therefore, we can state that for this very examined winter premium diesel fuel, calculated dielectric constant above about 6 may indicate that it may not be safe for engine. Future studies about other substances added should answer the question whether such value always corresponds to hazardous diesel, which should attract special attention. Until then, we believe that sporadic fake alarms would be better than a possible expensive repair of the modern fuel injection system. The greatest advantage of using impedance spectroscopy is that examination of matter can be performed without changing its properties and composition. This ability makes it possible to build a device that could work in situ. After proper calculations, it could warn a user that contamination was found, and examined diesel fuel can be potentially dangerous to the engine.

Acknowledgement: The authors would like to thank Professor Andreas Slemeyer from Technische Hochschule Mittelhessen University of Applied Sciences, Giessen, Germany for his meaningful suggestions, support and contribution to our studies on water content in diesel fuel using impedance spectroscopy.

5 References

[1] Agency of Toxic Substances and Disease Registry, "Toxicological profile for fuel oils," U.S. Department of Health and Human Services, Public Health Service, Atlanta, 1995.
[2] J. V. Gerpen, E. Hammond, L. Yu, et al., "Determining the influence of contaminants on biodiesel properties," SAE Technical Paper No. 971685, Warrendale, MI: SAE, 1997.
[3] J. D. Souza, M. Scherer, J. Cáceres, et al., "A close dielectric spectroscopic analysis of diesel/biodiesel blends and potential dielectric approaches for biodiesel content assessment," Fuel, vol. 105, pp. 705–710, 2013.
[4] A. Setayeshmehr, I. Fofana, C. Eichler, et al., "Dielectric spectroscopic measurements on transformer oil-paper insulation under controlled laboratory conditions," IEEE Transactions on Dielectrics and Electrical Insulation, vol. 15, no. 4, pp. 1100–1111, 2008.
[5] Ł. Macioszek and R. Rybski, "Evaluation of water content in diesel fuel using impedance spectroscopy," in XXI IMEKO World Congress Measurement in Research and Industry, Prague, Czech Republic: Czech Technical University, 2015, pp. 1799–1802.
[6] Oil and Gas Institute, "Laboratory results no. 37/ta2/2015: testing water content in the supplied specimens of diesel fuel," National Research Institute, Kraków, Poland, 2015.
[7] N. G. McCrum, B. E. Read, and G. Williams, Anelastic and Dielectric Effects in Polymeric Solids. New York: Wiley, 1967, pp. 108–111.
[8] H. Ming and L. Toh-Ming, Metal–Dielectric Diffusion Processes: Fundamentals. New York: Springer, 2012, pp. 11–20.
[9] W. C. Heerens and F. C. Vermeulen, "Capacitance of kelvin guard-ring capacitors with modified edge geometry," Journal of Applied Physics, vol. 46, no. 6, pp. 2486–2490, 1975.

Christian Weber, Markus Tahedl and Olfa Kanoun

A Novel Method for Capacitive Determination of the Overall Resistance of an Aqueous Solution

Abstract: In industrial applications, non-contacting capacitive sensors are used to detect conductive fluids in non-conductive containers. During use, layers of the material to be detected may stick to the inside of the container, leading to measurement deviations. It is therefore important to characterize the electrode impedance in dependence on frequency, conductivity and layer thickness of the medium. To characterize material properties, data are obtained and analysed over a broad frequency range. Investigations have shown that in order to characterize highly conductive media, very high measurement frequencies and bandwidth are required. In this contribution, we propose to characterize thin layers of aqueous solutions by utilizing only the low-frequency part of a spectrum. Compared to previously proposed methods, the new simple approach can characterize solutions with a conductivity of up to 90 $\frac{mS}{cm}$ using measurement frequencies from 1–3 MHz. The obtained overall resistance can be used to differentiate between a conductive build-up and the actual fill level.

Keywords: Impedance sensors, capacitive sensors, fill level, behavioral model

1 Introduction

Limit levels of a non-conductive container filled with a conductive fluid can be detected using capacitive sensors. In some applications, the sensors are mounted on the outside of a non-conductive container. Conductive films on the inside of the container influence the sensors operation and in some cases produce a false positive signal.

Many attempts to measure the fill level of conductive and non-conductive fluids have been reported. An overview of measurement techniques for mostly non-conductive fluids is given in [1] and [2]. In [3], a sensor for measuring the fill level of a conductive fluid is described. The primary electrode of the sensor is a non-inductively wound short-circuited coil around a non-conducting pipe. The fluid acts as a second electrode of a cylindrical capacitor. However, the influence of thin conductive layers in the vicinity of the primary electrode has not been reported.

Christian Weber and Markus Tahedl, ifm efector gmbh, Department of Technical Basics, IFM-Str. 1, 88069 Tettnang, Germany
Olfa Kanoun, Chair for Measurement and Sensor Technology, Technische Universität Chemnitz, Chemnitz, Germany

In a previous contribution [4], we proposed to extract material properties by fitting complex impedance spectra to electrical equivalent circuits (EECs). However, using EECs requires high measurement bandwidth to identify all unknown parameters [5]. Also, high conductivities cause very small time constants, which in turn require high measurement frequencies. This leads to limitations regarding maximum conductivity of the fluid to be characterized. Measurement frequencies up to 30 MHz had to be used to characterize solutions with a relatively low conductivity of 2.17 $\frac{mS}{cm}$.

Other researches [6] have proposed a method for contactless conductivity measurement of droplets in a segmented flow in a frequency range from 1 to 10 MHz. The proposed measurement setup is optimized in such a way that the coupling capacitance is much larger than the solution capacitance, which allows for the solution resistance to be directly calculated from the measured real part of the impedance spectrum. However, this is not always the case in an industrial application, where container walls may be much thicker. Containers may also be fabricated using lossy dielectrics, which introduces an additional influence in the real part of the impedance spectrum.

In this contribution, instead of using EECs, we propose a method that utilizes only the low-frequency part of the complex impedance spectrum, where the coupling constant phase element is dominant. We determine the loss factor and the intersection point of the complex impedance curve with the real axis using a linear fit to a first-order polynomial. The intersection point may then be used to characterize a given aqueous solution with respect to its layer thickness. The method is first tested with synthetic data using an analytical model [7]. The measurement effect is then validated by experiments.

2 Proposed Measurement Method

In impedance spectroscopy, one possibility for signal processing is to fit the obtained complex impedance spectra to a behavioural equivalent circuit model like in Fig. 1. The capacitive coupling into the solution through a lossy dielectric (i.e., polyoxymethylene (POM), polyethylene terephthalate (PET), etc.) can be represented by a constant phase element (CPE) while the medium can be represented by a series of N RC-elements. The time constant of the i-th RC-element is $\tau_i = R_i C_i$ [4, 7].

However, a high conductivity of the solution leads to small values of the resistances, which leads to small time constants and therefore to high measurement frequency and bandwidth requirements. To sufficiently characterize aqueous solutions with a high conductivity, we propose to only utilize the low-frequency part of the impedance spectrum Z, which is dominated by the coupling CPE and the sum of the N resistances.

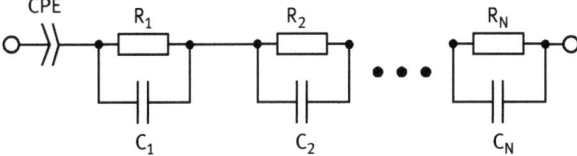

Fig. 1. Behavioural equivalent circuit model.

To determine the overall resistance $R_O \approx \sum_{i=1}^{N} R_i$ from that part, we need to fit the imaginary part of the impedance $\Im\{Z\}$ to the following first-order polynomial:

$$\Re\{Z\} = D \cdot \Im\{Z\} + R_O, \tag{1}$$

$\Re\{Z\}$ is the real part of the complex impedance, D is related to the loss factor and R_O is the intercept point of the polynomial and the Nyquist plot real axis. R_O and D are then determined by linear regression.

3 Impedance Simulation

In this section, synthetic impedance data are generated to test the proposed measurement method for different layer thicknesses and conductivities. Impedance data were calculated using a rotationally symmetric analytical model shown in Fig. 2. It consists of three different regions in which the conductivity σ and the permittivity ϵ may be

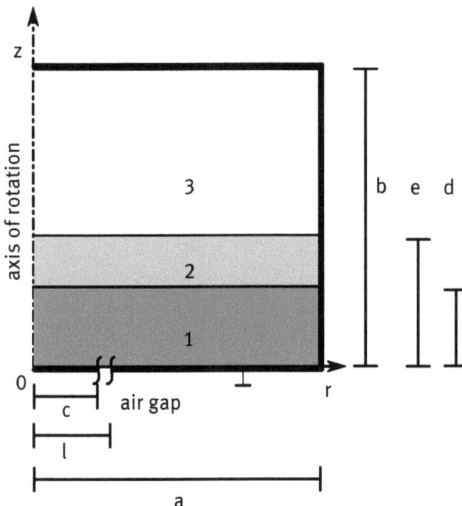

Fig. 2. Geometry of the rotationally symmetric analytical model.

freely chosen. Typically, region one models the lossy dielectric of the container wall, region two models the conductive solution and region three is air. The dimensions a, b, c, d, e and l are variable. Dirichlet boundary conditions are employed to model the measurement electrode from 0 to c. Impedance is calculated by integrating the current density across the electrode surface area. A more detailed description is given in [7].

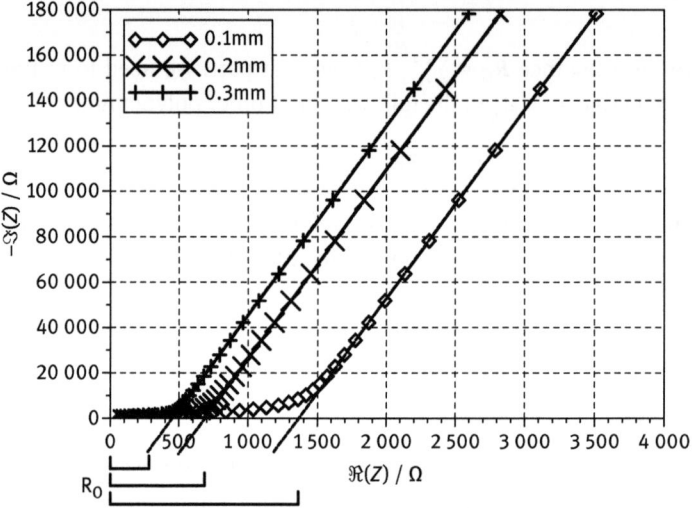

Fig. 3. Nyquist plot of the modelled impedance for several layer thicknesses. R_O is the intersection point of the extrapolated low-frequency data with the real axis.

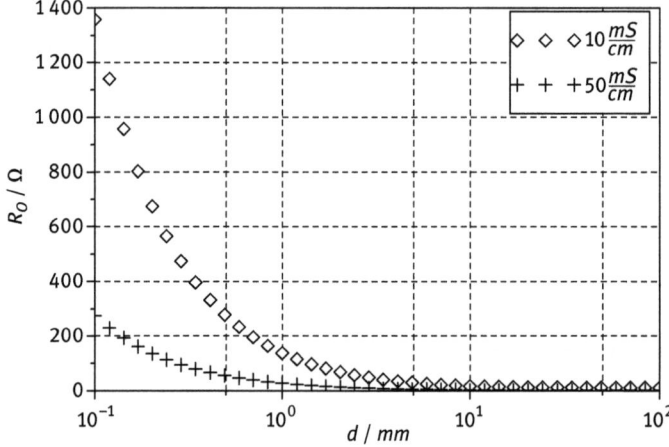

Fig. 4. Overall resistance R_O versus layer thickness d_l. R_O allows for a clear discernibility of layers thicknesses between 0.1 and 5 mm.

Tab. 1. Fitted values for a constant geometry and two conductivities

Conductivity $\left[\frac{mS}{cm}\right]$	D	R_O [Ω]
1	0.01223	11260
10	0.012	1139
100	0.012	113.9
1000	0.012	11.39

R_O was determined in a frequency range from 100 to 640 kHz. In this range, the complex impedance curve can in good approximation be described as a first-order polynomial for the given geometry. Fig. 3 illustrates the method for determining R_O, which depends on the layer thickness for a given conductivity. Fig. 4 shows R_O versus several layer thicknesses $d_l = e - d$. R_O is related to the conductivity of the fluid and the coupling CPE as can be seen in Tab. 1.

If the conductivity is sufficiently high, D approaches a constant value, which is equal to the dissipation factor of the lossy dielectric in region one of the model, because a sufficiently conductive layer acts as the second electrode of a parallel plate capacitor. Once D has reached a constant value, R_O is proportional to the conductivity of the fluid because of the constant geometry. It can be seen from Fig. 4 that R_O depends on the layer thickness of an aqueous solution for a given conductivity. Dynamic range and sensitivity are dependent on the conductivity of the solution.

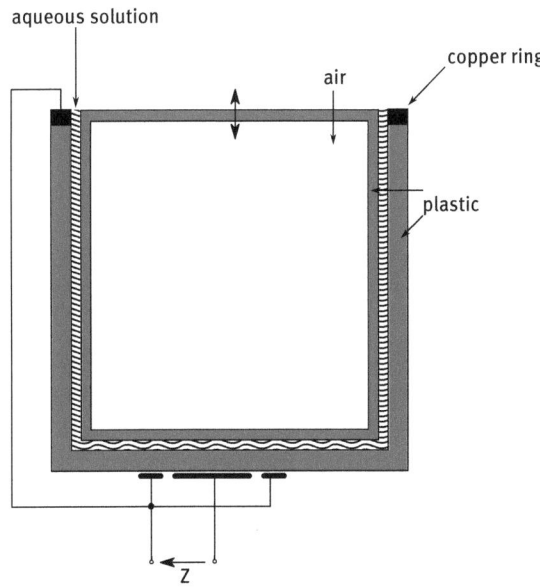

Fig. 5. Sketch of the measurement setup.

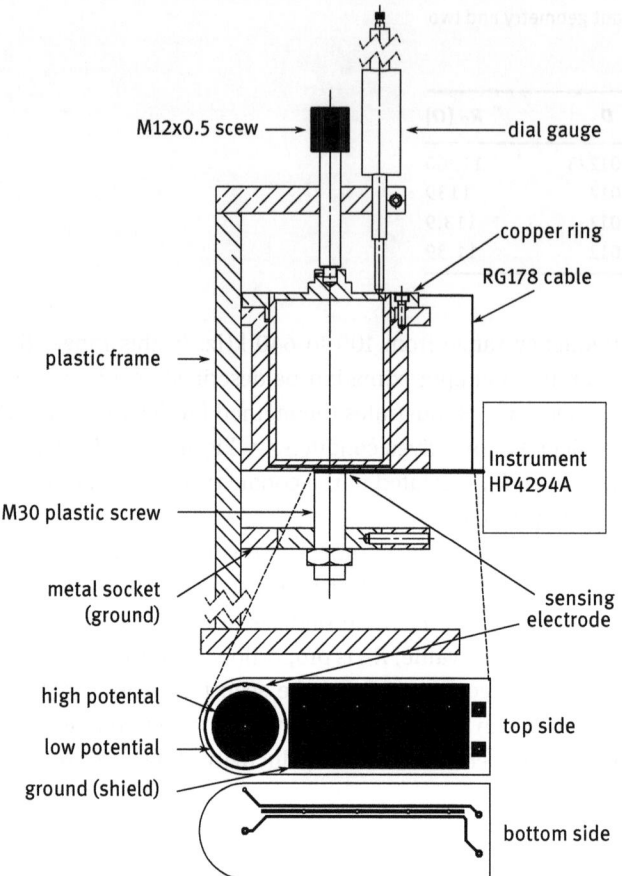

Fig. 6. Detailed design of the measurement setup.

4 Experimental Verification

In this section, the method is verified using an experimental setup shown in Figs. 5 and 6.

This setup can be used to produce thin layers of aqueous solutions and measuring their impedance in a frequency range from 20 kHz to 30 MHz [7]. The setup consists of an inner and an outer cylinder fabricated from plastic. The height of the inner cylinder is adjustable, which leads to a gap with variable thickness. The gap is filled with an aqueous solution of known conductivity. A copper ring is used to ground the solution and ensure a defined current path. A commercial impedance analyser (HP4294A, Keysight Technologies, Santa Rosa, CA) is used to measure the impedance between the sensing electrode and the instruments low potential port.

Potassium chloride solution was used in the measurements because its conductivity is well defined. A more detailed description is available in [7]. Measured data have been checked for consistency using a modified linear Kramers-Kronig fitting algorithm [7–9].

Fig. 7 shows measured and consistency-checked data in a frequency range from 1 to 3 MHz for a solution conductivity of $9.53\,\frac{mS}{cm}$ at a temperature of $27.1°C$. It can be seen that the overall resistance decreases monotonously with increasing layer thickness, as was shown in Section 3.

Tab. 2 shows R_O values fitted from the measured and consistency-checked data for two different conductivities. There is sufficient dynamic range to distinguish different layer thicknesses even for very high conductivities. This is a significant improvement compared with the methods previously proposed.

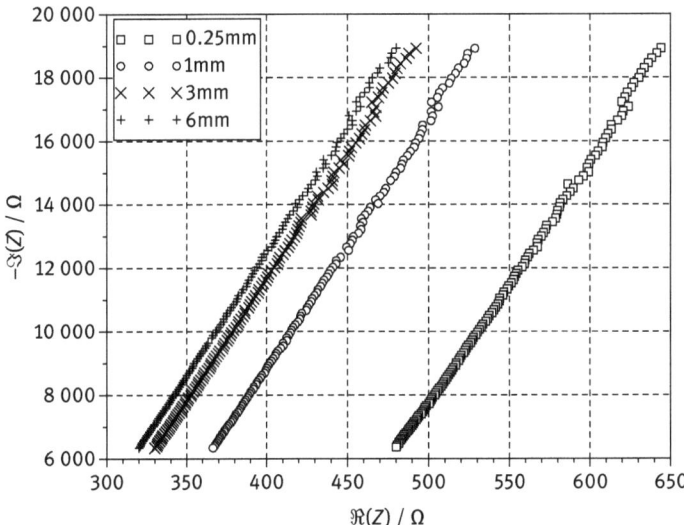

Fig. 7. Measured and consistency-checked data in a frequency range from 1 to 3 MHz for several layer thicknesses. KCl solution $9.53\,\frac{mS}{cm}$ at $27.1°C$.

Tab. 2. Fitted R_O values for measured and validated data and two different conductivities

d_l [mm]	$9.53\,\frac{mS}{cm}$	$90.38\,\frac{mS}{cm}$
0.25	400	39
1	287	31
3	251	27
6	241	25

5 Conclusion

A method for characterizing the overall resistance of conductive aqueous solutions using capacitive coupling utilizing only the low-frequency part of the complex impedance spectrum has been proposed. It allows for characterization of solutions with a significantly higher conductivity while maintaining relatively low measurement frequencies. The method and the measurement effect were described using synthetic impedance spectra calculated by an analytical model. It was shown in an experimental investigation that the modelled measurement effect is in qualitative agreement with the measured effect, even though a quantitative agreement could not be reached. With the proposed method, it is possible to implement a non-contacting capacitive sensor for detecting conductive fluids, which is robust against thin conducting layers in vicinity of the sensing electrode.

6 References

[1] H. Zangl, A. Fuchs, and T. Bretterklieber, "Non-invasive measurements of fluids by means of capacitive sensors," Elektrotechnik und Informationstechnik, vol. 126, no. 1–2, pp. 8–12, 2009.

[2] B. Kumar, G. Rajita, and N. Mandal, "A review on capacitive-type sensor for measurement of height of liquid level," Measurement and Control, vol. 47, no. 7, pp. 219–224, 2014 (eprint: http://mac.sagepub.com/content/47/7/219.full.pdf+html).

[3] S. Bera, J. Ray, and S. Chattopadhyay, "A low-cost noncontact capacitance-type level transducer for a conducting liquid," IEEE Transactions on Instrumentation and Measurement, vol. 55, no. 3, pp. 778–786, 2006.

[4] C. Weber, M. Tahedl, and O. Kanoun, "Capacitive measurement setup for characterizing thin layers of aqueous solutions," in International Workshop on Impedance Spectroscopy, Chemnitz, Germany, 2014.

[5] O. Kanoun, U. Tröltzsch, and H.-R. Tränkler, "Benefits of evolutionary strategy in modeling of impedance spectra," Electrochimica Acta, vol. 51, no. 8–9, pp. 1453–1461, 2006.

[6] B. P. Cahill, R. Land, T. Nacke, et al., "Contactless sensing of the conductivity of aqueous droplets in segmented flow," Sensors and Actuators B: Chemical, vol. 159, no. 1, pp. 286–293, 2011.

[7] C. Weber, F. Wendler, M. Tahedl, et al., "2D-Modellierung des Einflusses leitfa ähiger Schichten auf die Sensorimpedanz bei kapazitiven Sensoren," TM - TECHNISCHES MESSEN, vol. 81, no. 9, pp. 442–449, 2014.

[8] M. Schönleber, D. Klotz, and E. Ivers-Tiffée, "A method for improving the robustness of linear Kramers-Kronig validity tests," Electrochimica Acta, vol. 131, pp. 20–27, 2014.

[9] U. Tröltzsch, "Modellbasierte Zustandsdiagnose von Geraetebatterien," PhD thesis, Universität der Bundeswehr München, 2005.

Part III: **Material Characterization**

Bernhard Roling and André Schirmeisen

Nanoscale Electrochemical Characterization of Materials by means of Electrostatic Force and Current Measurements

Abstract: Atomic force microscopy (AFM) techniques are well suited for measuring nanoscale electrical and electrochemical properties of materials. By using conductive AFM tips as nanoelectrodes, different physical quantities can be measured, in particular electrostatic forces, currents and mechanical strains. The spatial resolution of these techniques is usually of the order of the tip diameter, that is in a range from 20 to 100 nm. Here, we describe the usage of time-domain electrostatic force spectroscopy (TD-EFS) and nanoscale impedance spectroscopy for obtaining information about ion transport processes in heterogeneous materials. We take partially crystallized glass ceramics as model materials for demonstrating that dynamic processes in the glassy phase, in crystallites and at interfaces can be distinguished via TD-EFS. Furthermore, by applying grid-type spectroscopic measurements, maps of the local TD-EFS relaxation strength were obtained, giving information about the spatial distribution of the glassy and crystalline phase. Finally, we show that nanoscale impedance spectroscopy during metal particle formation on solid-ion conductors can be used for measuring local ionic conductivites.

Keywords: Time-domain electrostatic force spectroscopy, nanoscale impedance spectroscopy

1 Introduction

Broadband impedance spectroscopy is one of the key tools for characterizing ion transport in solids and liquids as well as for characterizing electrochemical processes [1–3]. In traditional macroscopic impedance set-ups, the electrode area is in the range of 1 cm^2 and consequently information averaged over this area is obtained. This is a major drawback when heterogeneous materials are under study, in which nanoscale structural features exert a strong influence on ion transport mechanisms [4–7].

To obtain information at the nanoscale, different electrical atomic force microscopy (AFM) techniques can be used. One technique is based on the measurements of electrostatic forces between a conductive tip and the sample without contact. Fig. 1 shows the principle set-up of an electrostatic force microscope. In this technique, the

Bernhard Roling, Fachbereich Chemie, Philipps-Universität Marburg, Marburg, Germany
André Schirmeisen, Institut für Angewandte Physik, Justus-Liebig-Universität Gießen, Gießen, Germany

DOI 10.1515/9783110449822-010

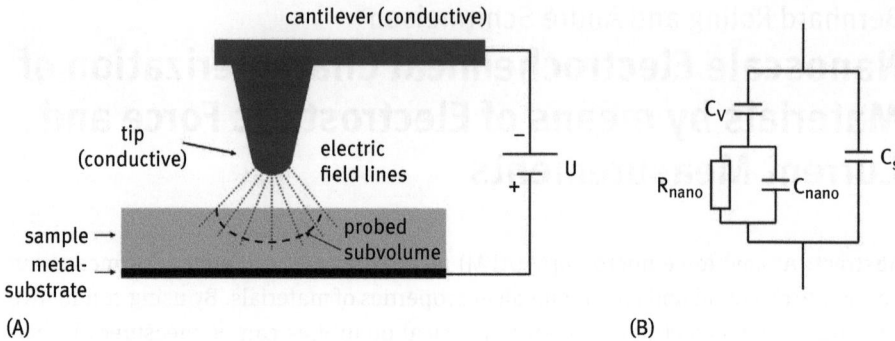

Fig. 1. (A) Schematic illustration of the experimental set-up for electrostatic force spectroscopy on solid electrolytes. (B) Equivalent circuit for modelling the overall capacitance of the system.

tip is located above the sample surface, and an electric potential is applied to the tip, while the sample is connected to a counter electrode. In this case, the electric field around the tip decays radially, so that the major part of the potential drop in the sample occurs within a small sub-volume underneath the tip, the sub-volume being of the order of the tip diameter cubed. This allows for measuring the electric sample properties with a local resolution of the order of the tip diameter. In contrast to conventional force microscopy, which is sensitive to surface properties, here the signal is sensitive to subsurface and bulk properties. Typical tip radii are of the order of 10–50 nm, which results in a local resolution of about 100 times better than methods using micro-electrodes [8].

Electrostatic force measurements as a function of time after applying the potential to the tip can be used to monitor ion transport processes in solid electrolytes. This technique is called 'time-domain electrostatic force spectroscopy' (TD-EFS). The tip of a conductive cantilever is oscillating with small amplitudes of 1–3 nm above the surface of the solid ionic conductor. If a negative potential is applied to the tip, the electrical field emanating from the tip penetrates the sample causing the positively charged ions to accumulate underneath the tip. In the case of small cantilever oscillations and long-ranged electrostatic interactions, the frequency shift induced by the tip bias potential ΔU is given by [9]:

$$\Delta f(t) = -\frac{f_0}{4c} \times U^2 \times \frac{\partial^2 C(t)}{\partial z^2} \qquad (1)$$

where $C(t)$ denotes the overall capacitance between biased tip and ground, c is the normal spring constant and f_0 is the free cantilever resonance frequency. $C(t)$ can be modelled by an equivalent circuit illustrated in Fig. 1B. The probed nanoscopic sub-volume of the sample is represented by a resistor R_{nano} in parallel to a capacitor C_{nano}. The resistor R_{nano} models ionic conduction, while C_{nano} models the capacitance because of electronic and vibrational polarization. The gap between tip and probed sub-volume is represented by a vacuum capacitor CV in series to the $R_{nano}C_{nano}$

element. Additionally, a capacitor CS in parallel to the other elements is introduced, which represents all stray capacitances between tip and ground. Upon application of the voltage U, all capacitors are instantaneously charged. Subsequently, the capacitor C_{nano} is discharged through the resistor R_{nano}. This leads to an increase of the overall capacitance $C(t)$ and thus to a decrease of the resonant frequency. In the framework of the equivalent circuit, the time dependence of $C(t)$ is given by [10]:

$$C(t) = C_V \left[1 - \frac{C_V}{C_{nano} + C_V} \times \exp(-t/\tau) \right] + C_S \qquad (2)$$

with $\tau = R_{nano} \times (C_{nano} + C_V)$. From a microscopic point of view, the discharge of the sample capacitor is due to mobile ions moving in the direction of the electric field, until the field in the probed sub-volume becomes zero. Thus, the time-dependent drop of the cantilever resonance frequency reflects the time-dependent built-up of an ionic double layer at the surface of the solid ionic conductor. The second possibility is the measurement of local ionic currents by coupling the AFM with an impedance spectrometer. For a tip in contact with a solid sample, O'Hayre et al. [11] proposed a simple equivalent-circuit model shown in Fig. 2. R_{Tip} denotes the resistance of the tip, which is expected to be low. R_{Cont} and C are the tip/sample contact resistance and the tip/sample capacitance, respectively. R_{Spread} is the spreading resistance related to the local dc conductivity of the sample via [8, 12]:

$$R_{Spread} = \frac{1}{4 \times r \times \sigma_{dc}} \qquad (3)$$

Here, r stands for the radius of curvature of the tip.

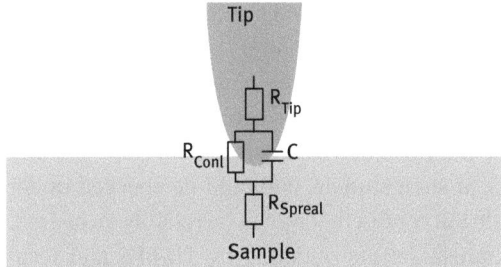

Fig. 2. Equivalent circuit for the sample / tip interface according to O'Hayre et al. [11].

In the case of a semiconducting sample, R_{Cont} represents the tip/sample contact resistance, while the capacitance C is determined by electronic space charge layers at this interface and possibly by geometrical non-idealities of the tip/sample contact. If R_{Cont} is much smaller than R_{Spread}, the measured current is determined by R_{Spread}, and thus information about the local electronic conductivity of the sample can be obtained. O'Hayre et al. [11] showed that by using high spring-constant cantilevers

(>10 N/m) and by applying contact forces larger than 0.5 μ N, the contact resistances can be reduced considerably, and local electrical properties of Au/Si_3N_4 test structures and of ZnO varistors can be measured.

In the case of solid ion conductors, R_{Cont} can be identified with the electrochemical charge transfer resistance at the tip/sample interface, while C represents the double-layer capacitance of this interface. When the electrochemical charge transfer resistance is high and the tip resistance is negligible, the circuit reduces to a spreading resistance R_{Spread} acting in series to a double-layer capacitance. Double-layer capacitances per unit area are typically in the range of 10 $\mu F/cm^2$. When we assume a tip radius of 50 nm and a double-layer area of πr^2, the absolute double-layer capacitance is as small as $C_{DL} \approx 5 \cdot 10^{-18} F$. In the case of a sample with a local conductivity of $\sigma_{dc} \approx 10^{-6} S/cm$, the spreading resistance is $R_{Spread} \approx 5 \cdot 10^{10} \Omega$, and the double-layer charging time is given by:

$$\tau_{DL} = R_{Spread} \times C_{DL} \approx 5 \cdot 10^{10} \Omega \cdot 5 \cdot 10^{-18} F \approx 2.5 \cdot 10^{-7} s \qquad (4)$$

This implies that at frequencies below $1/\tau_{DL} \approx 50$ MHz, the impedance spectra are governed by double-layer formation (electrode polarization), making it virtually impossible to obtain information about the spreading resistance and thus about the local ion conductivity. This is unfavourable, since the typical frequency range of impedance measurements extends from the sub-Hz regime to a few MHz.

The only way to get rid of this problem is the short-circuiting of the double-layer capacitance by a low charge transfer resistance. This can be done by applying a dc bias voltage, thus providing a sufficient over-potential for a Faradaic process. One example is a study of O'Hayre et al. [13] on dry and hydrated Nafion membranes for fuel cell applications. They were able to measure the impedance of charge transfer reactions at the tip/membrane interface, when they applied a cathodic bias to the tip while supplying hydrogen gas at the counter electrode (anode).

Considering the relevance of nanoscale characterization of ion conducting materials for modern electrochemical cells, it seems important to carry out comprehensive studies of processes taking place at the interface between conducting AFM tips and solid ion conductors. In the framework of such studies, it should be checked under what experimental conditions nanoscale ion conductivities can be reliably measured. In the following, we present representative results obtained by TD-EFS and local impedance measurements.

2 Experimental

For the TD-EFS experiments, we used a variable-temperature AFM operating under ultrahigh vacuum (UHV) conditions (Omicron VT-AFM). The sample temperature could be varied in a range from 30 up to 650 K. The force sensor was a single crystalline, highly doped silicon cantilever with a resonant frequency of 300 kHz and a spring constant of 20 N/m, featuring a sharp conducting tip with an apex radius of 10 nm (non-contact cantilever of type NCHR from Nanosensors). The system was operated in the frequency modulation-mode (FM-mode), where the cantilever is always oscillating at the resonant frequency. Conservative tip-sample forces induce a shift in the resonant frequency, which is used as the feedback parameter for the tip-sample distance control during surface scanning. During the TD-EFS measurements, we applied a negative bias voltage to the tip, while the counter electrode was kept at ground potential, and we measured the time-dependent electrostatic forces between tip and sample caused by local movements of ions.

Nanoscale impedance spectroscopy was performed under ambient conditions using a NT-MDT-Solver P47 atomic force microscope with a conductive diamond tip acting as working electrode. An ion-blocking platinum layer was used as large-area counter electrode. In this set-up, the main potential drop is concentrated at the AFM tip, so that a third reference electrode is not needed. To obtain reproducible results, the usage of cantilevers with high spring constants in the range of 40 N/m was necessary. We chose conductive diamond-coated cantilevers (Nanosensors GmbH, CDT-NCHR). The electrochemical measurements were performed in contact mode with tip/sample forces in the range of 1–2 μN. For the chronoamperometric measurements, the AFM was connected to a Novocontrol Alpha-AK impedance analyzer equipped with a POT/GAL 15V/10A electrochemical interface.

3 Results and Discussion

3.1 Time-Domain Electrostatic Force Spectroscopy (TD-EFS)

TD-EFS measurements were carried out on glass ceramics of composition $LiAlSiO_4$ with different degrees of crystallinity. The Li+ ion transport mechanisms in these glass ceramics depend on the degree of crystallinity, since $LiAlSiO_4$ glass is a moderate ion conductor with an activation energy of E_A^{glass} = 0.72 eV, while the crystalline $LiAlSiO_4$ phase is a poor ion conductor with a high activation energy of $E_A^{crystal}$ = 1.07 eV [14].

When TD-EFS measurements are carried out on $LiAlSiO_4$ glass ceramics, the probed sub-volume contains, depending on the position of the tip, different amounts of glassy phase, crystallites and interfacial areas. At different temperatures, the contribution from the different phases varies strongly, since their relaxation times $\tau = R_{nano} \cdot (C_{nano} + C_V)$ exhibit different activation energies. The time scale of TD-EFS measurements ranges typically from 1 ms to 10 s. Consequently, the sample temperature has to be adjusted in a way that the relaxation time of one phase is in this time window, so that the contribution of this phase dominates the measured signal. For example, at room temperature, movements of ions in the glassy phase govern the TD-EFS signal, while the ions in the crystalline phase are immobile on the experimentally accessible time scales [15]. On the other hand, at elevated temperatures, the ions in the crystallites will start to contribute significantly, while the relaxation of the ions in the glassy phase becomes too fast to be resolved.

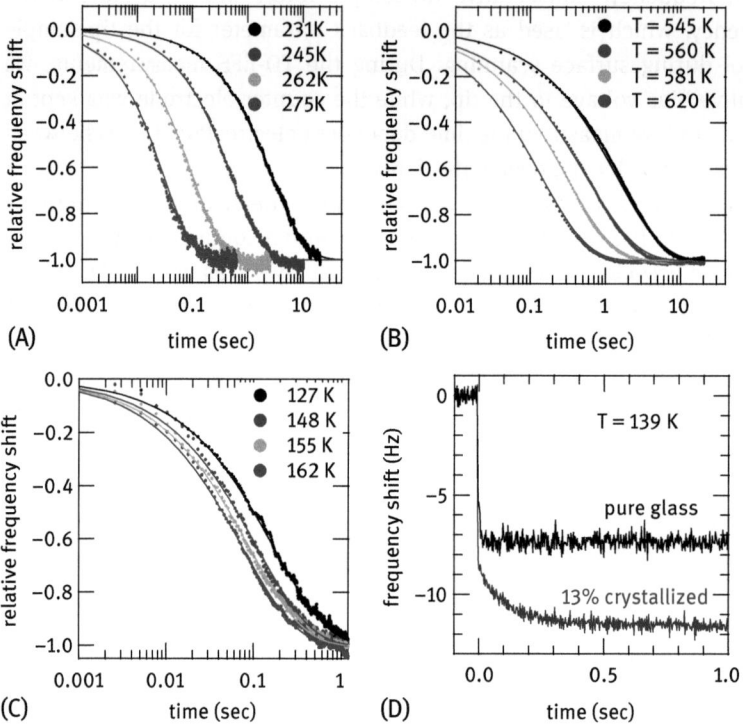

Fig. 3. TD-EFS relaxation curves of an LiAlSiO$_4$ glass ceramic with 13% crystallinity in three different temperature regimes. The frequency shift axis was normalized to unity for a better comparison of the relaxation times. The inset compares raw data curves obtained at low temperatures for the glass ceramic and for a pure glass sample, proving an additional ion transport process in the glass ceramic.

In Fig. 3, we show TD-EFS relaxation curves of the cantilever resonance frequency for an LiAlSiO$_4$ glass ceramic with 13% crystallinity. For a better comparison of curves obtained at different temperatures, the frequency shifts were normalized to unity. The thermally activated ionic movement in the different phases leads to a decrease of the relaxation time τ with increasing temperature T. The curves in Fig. 3A in a temperature range from 231 to 275 K reflect ionic movements in the glassy phase, while the curves in Fig. 3B in a temperature range from 545 to 620 K are governed by movements in the crystalline phase. Fig. 3C shows relaxation curves obtained in a temperature range from 127 to 162 K. In this temperature range, we detect an additional relaxation process, which is most likely due to ion transport at the interface between crystallites and glassy phase. Similar to the relaxation curves at higher temperatures we find a systematic shift of the curves to the left with increasing temperature, indicative of a thermally activated ion hopping process. To exclude possible artefacts we performed additional test measurements on a homogeneous glass sample without internal interfaces. The inset in Fig. 3C shows representative relaxation curves of the pure glass sample and of the glass ceramic with 13% crystallinity in direct comparison at the same temperature $T = 139$ K. Only the partially crystallized sample shows a clear relaxation process at this temperature.

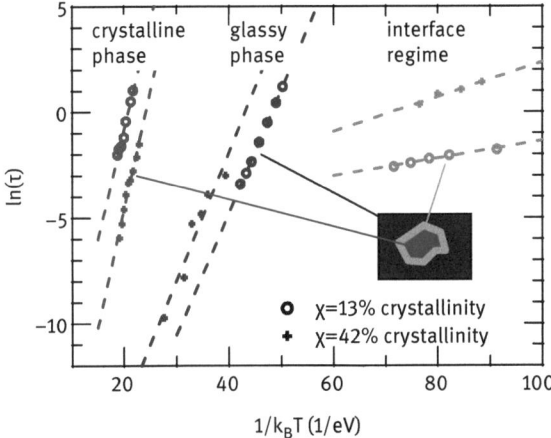

Fig. 4. Arrhenius plot of the TD-EFS relaxation times for LiAlSiO$_4$ glass ceramic with 13% and 42% crystallinity, respectively. For both samples, three relaxation processes can be identified. For comparison, the solid lines denote the macroscopic relaxation times for a pure glass sample and for a completely crystallized sample, respectively.

In the Arrhenius plot in Fig. 4, we show a compilation of the TD-EFS relaxation times for two different glass ceramics with 13% and 42% crystallinity, respectively. We can identify three distinct regimes: around room temperature, a fit with the Arrhenius law yields an activation energy of 0.58 and 0.61 eV for the samples with 13% and 42% crystallinity, respectively (see grey markers [data] and dashed grey lines [fit] in Fig. 4). For comparison, the solid black line denotes the corresponding relaxation times from macroscopic impedance measurements on a pure glass sample with an activation energy of 0.71 eV. Within the experimental error, which is mainly due to uncertainties in the sample temperature during the TD-EFS measurements, the nanoscopic and the macroscopic results are in reasonable agreement. At temperatures above 500 K, the TD-EFS relaxation processes exhibit activation energies of 1.03 and 1.11 eV for the two samples with 13% and 42% crystallinity, respectively (see blue markers [data] and dashed blue lines [fit] in Fig. 4). These activation energies are very close to the activation energy of 1.07 eV determined from macroscopic conductivity measurements of a completely crystallized $LiAlSiO_4$ ceramic (solid dark blue line in Fig. 4). Remarkably, the TD-EFS relaxation times of the glass ceramics deviate in a systematic fashion from the macroscopic electrical relaxation times. In Refs. [15, 16] we offered a qualitative explanation for this phenomenon.

Below room temperature, the Arrhenius fit yields an activation energy of 0.04 and 0.08 eV for the two investigated samples (see red markers [data] and dashed red lines [fit] in Fig. 4), which indicate ionic movements at the interfaces with activation energies of only a few times $k_B T$ [16]. Since this activation energy is much lower than the activation energy for macroscopic ion transport (about 0.7 eV), we conclude that the interfacial areas do not form percolating pathways across the glass ceramic. Consequently, the higher activation energy for ion transport across the glassy phase governs the activation energy for macroscopic transport. On the other hand, the interfacial regions act as local electrical shorts, thus leading to a moderate but significant increase of the ionic conductivity compared with a pure $LiAlSiO_4$ glass.

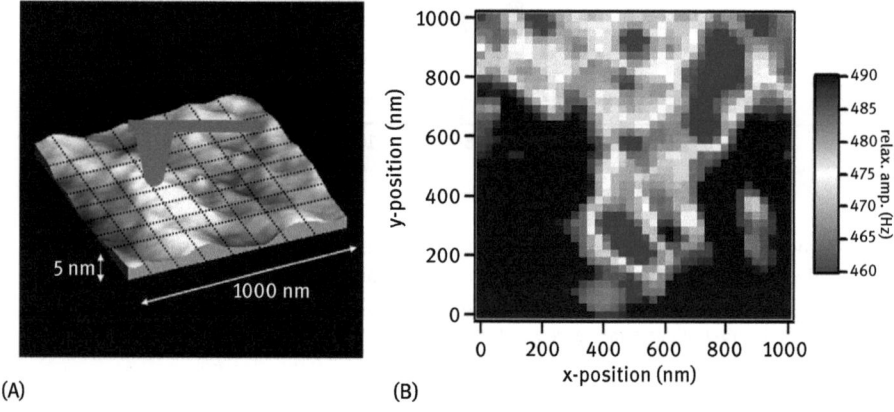

(A)   (B)

Fig. 5. (A) Schematic of TD-EFS grid measurements, where a time-dependent relaxation curve is obtained at each point of a pre-defined grid on a previously scanned surface topography. (B) Experimental data from a TD-EFS grid measurement with a color coding of the relaxation strength. Red areas denote positions with lower relaxation strength, indicating a larger amount of crystalline phase in the probed subvolume.

In order to directly visualize differences in the ion transport in different phases, we combine the technique of grid spectroscopy with the TD-EFS method. As shown in the sketch in Fig. 5A, an equally spaced grid is predefined on the surface of a glass ceramic, and at each position the time evolution of the frequency shift signal is acquired. In general, the signal will be a volume-averaged mixture of the two major phases (glassy and crystalline) within the probed sub-volume (here we neglect the comparably minute contributions from the interfacial areas). Since at room temperature, the ions in the crystalline phase are too slow to contribute to the TD-EFS signal, the relaxation strength should increase with increasing volume fraction of the glassy phase in the probed subvolume. A corresponding measurement is shown in Fig. 5B on a glass ceramic with 42% crystallinity. The 1,000 x 1,000 nm surface area was divided into a 40 x 40 grid of 1,600 spectroscopy points. At each position of the tip, the relaxation strength is shown, color-coded from red (lower relaxation strength) to blue (higher relaxation strength). The positions of lower relaxation strength are coherent red areas, indicating the presence of a large amount of crystalline phase. The resulting picture in Fig. 5B suggest that crystallites with diameters in the range from 100 to 200 nm exist in the glassy matrix, in good agreement with the results of transmission electron microscopy studies [15]. Thus in the case of nanomaterials, which exhibit a close relation between structural and dynamic heterogeneities, TD-EFS turns out to be a valuable technique for obtaining structural information on nanoscopic length scales.

3.2 Nanoscale Impedance Spectroscopy

We have taken silver phosphate glass samples of composition 20 AgI–80 AgPO$_3$ (glass I-20), 10 AgI–90 AgPO$_3$ (glass I-10) and AgPO$_3$ (glass I-0) as model materials for demonstrating a protocol suitable for measuring local ion conductivities [17, 18]. A Pt electrode (area 0.04 cm^2), acting as counter electrode, was sputtered onto the same face of the sample that was subsequently studied with the AFM tip. The nanoscopic impedance measurements were performed using a NT-MDT Solver P47 atomic force microscope with a conductive tip acting as working electrode. The AFM was connected to the Alpha-AK impedance analyzer equipped with a POT/GAL 15V/10A electrochemical interface, see Fig. 6. In order to carry out reproducible current and impedance measurements, the usage of cantilevers with high spring constants in the range of 40 N/m were necessary. We chose conductive diamond-coated cantilevers

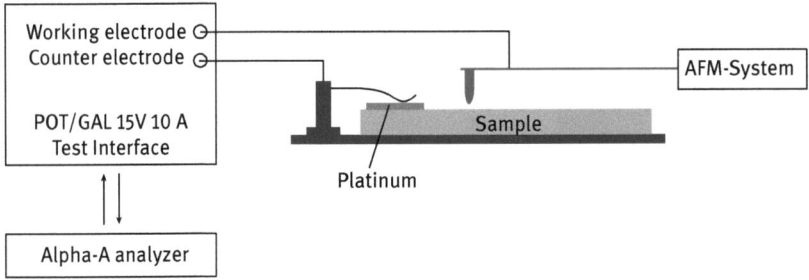

Fig. 6. Schematic illustration of the set-up for measuring local ionic currents by means of nanoscale impedance spectroscopy. A Pt electrode acting as counter electrode was sputtered onto the same face of the sample that was investigated by the AFM tip.

(Nanosensors GmbH, CDT-NCHR). The impedance measurements were performed in contact mode with tip/sample forces in the range of 1–2 μ N. The same set-up was used to carry out cyclic voltammetry measurements.

In Fig. 7, we show, as an example, a cyclic voltammogram of the I-20 glass obtained at a scan rate of 10 V/s. When scanning the tip potential from 0 V into the cathodic regime, the following characteristics were observed: (i) A critical potential was necessary to detect a current response. (ii) Above the critical potential, the current increases strongly with increasing cathodic potential. These characteristics can be explained as follows: With increasing cathodic potential, the overpotential for Ag^+ ion reduction increases, leading to a strong increase of the Ag^+ reduction current and the

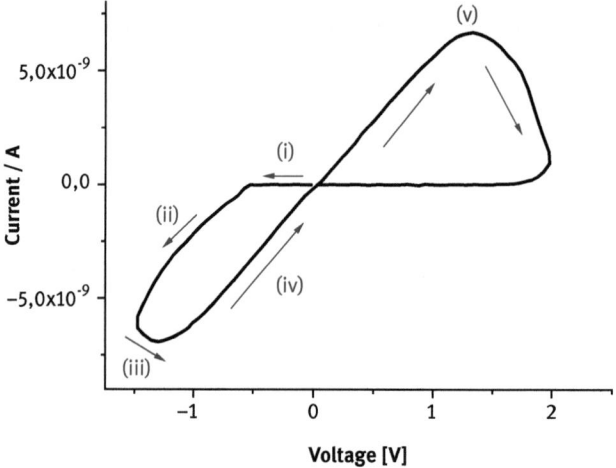

Fig. 7. Nanoscopic cyclic voltammogram of the glass I-20 taken with a scan rate of 10 V/s.

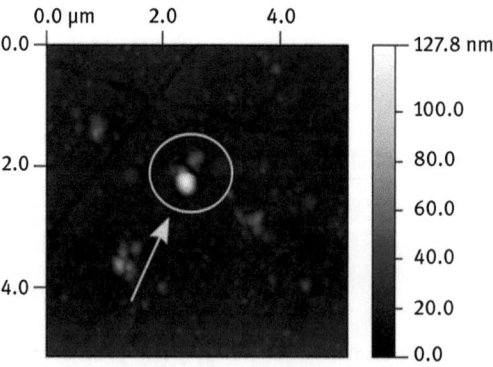

Fig. 8. Topographic image of a silver metal nanoparticle formed on the surface of the glass I-20.

formation of Ag nanoparticles on the surface of the glass. Fig. 8 shows a topographic image of a typical Ag nanoparticle on the surface of the glass I-20.

The current/potential characteristics for the backward scan to 0 V are quite remarkable: (iii) Starting around −1.5 V, the current increases further and shows a weak maximum around −1.3 V. (iv) At lower cathodic potentials, the current/voltage relation becomes, to a good approximation, linear. These characteristics can be rationalized by the following assumptions: In regime (iii), the lateral growth of the silver particle leads to an increasing current, even though the scan direction has been reversed. In regime (iv), there is almost no further lateral growth, and the current is limited by the spreading resistance of the sample. (v) During the anodic scan, an anodic peak is observed pointing to an oxidation process.

The silver particle growth and the experimental conditions for measuring the spreading ionic resistance of the samples were further investigated by ac impedance spectroscopy. The most suitable measurement protocol was the following: We applied an ac voltage with constant frequency and amplitude and measured the real part of the nanoscopic admittance Y_{nano} as a function of time. In Fig. 9, results for the I-0 glass are shown, which were obtained at different tip positions on the surface. In this case, the chosen frequency and rms ac voltage were 1 Hz and 4 V, respectively. Within the first minute, the admittance Y_{nano} increased with increasing time indicating lateral particle growth. In this time regime, the results obtained at different tip positions were in close agreement. At longer times, the admittance levelled off into a plateau regime indicating no further lateral particle growth. For comparison, the sizes of the Ag particles were studied by topographic imaging (as shown in Fig. 8). The lateral radius of the particles, r, was determined and used for normalizing the admittance Y_{nano} according to the spreading resistance relation (1).

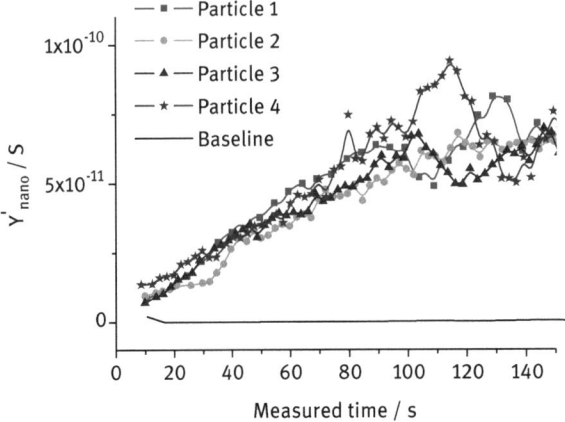

Fig. 9. Real part of the nanoscopic admittance Y_{nano}, of the glass I-0 measured at a given frequency of 1 Hz and a given rms ac voltage of 4 V plotted versus measurements time.

In Fig. 10, we plot the nanoscopic $Y_{nano}/4r$ data versus the macroscopic dc conductivity of the glasses, the latter measured by conventional impedance spectroscopy. The $Y_{nano}/4r$ data points were obtained at different tip positions above the surface. It is obvious that the data points scatter around the line $Y_{nano}/4r = \sigma_{dc}^{macro}$. We found that the amount of scattering depends on the roughness of the surface. The lowest roughness was achieved for the I-20 glass. For this glass, the scattering of the $Y_{nano}/4r$ values around σ_{dc}^{macro} is not larger than a factor of 2. For the other glasses with rougher surfaces, this factor was in the range of 3–4.

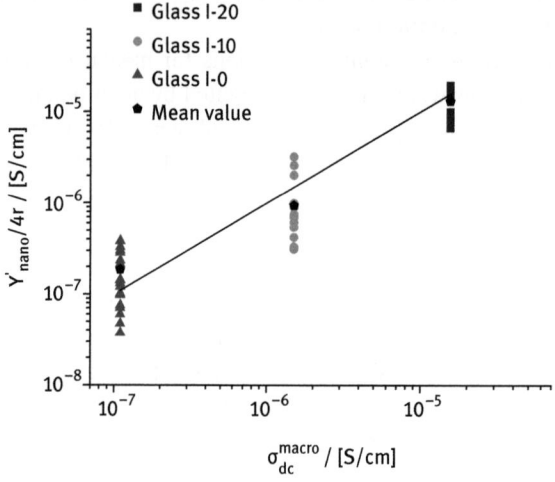

Fig. 10. Plot of the nanoscopic $Y_{nano}/4r$ data versus the macroscopic dc conductivity σ_{dc}^{macro} of the I-20 (blue squares), I-10 (green dots) and the I-0 (red triangles) glass. The mean values of $Y_{nano}/4r$ are symbolized by the black pentagons. The solid line is given by $Y_{nano}/4r = \sigma_{dc}^{macro}$.

Averaging of the $Y_{nano}/4r$ values obtained for a given glass at different tip positions resulted in mean values (black pentagons in Fig. 9), which differed only by 15–70% from σ_{dc}^{macro}. This demonstrates that that our measurement protocol is well suited for obtaining reliable mean values for the spreading resistance and thus the local ionic conductivity.

4 Conclusion

We have shown that TD-EFS is capable of differentiating between ion transport processes in different phases (crystallites, amorphous phase, interface between crystallites and amorphous phase) of Li^+ ion conductive $LiAlSiO_4$ glass ceramics.

The activation energies of the processes could be determined separately by carrying out TD-EFS measurements in different temperatures and therefore relaxation time regimes. By combining TD-EFS with grid spectroscopy, the distribution and size of the crystallites could be visualized.

Furthermore, we have demonstrated a measurement protocol for obtaining local ion conductivities of glasses by means of nanoscale impedance spectroscopy. Since in the nanoscale geometry, double-layer formation is generally very fast, the double-layer capacitance at the ionic conductor/tip interface has to be short-circuited by a low charge transfer resistance for ion reduction. That means that during the reductive formation of metal particles on the surface of the ion conductor, a simultaneous measurement of ionic current and metal particle size can be used to calculate local ion conductivities via the spreading resistance formula. The suitability of the measurement protocol was proven on the basis of AgI–AgPO$_3$ glasses with different compositions and macroscopic Ag$^+$ ion conductivities.

5 References

[1] F. Wohde, M. Balabajew, and B. Roling, "Li+ transference numbers in liquid electrolytes obtained by very-low-frequency impedance spectroscopy at variable electrode distances," Journal of the Electrochemical Society, vol. 163, pp. A714–A721, 2016.

[2] R. Atkin, N. Borisenko, M. Drüschler, et al., "Structure and dynamics of the interfacial layer between ionic liquid and electrode materials," Journal of Molecular Liquids, vol. 192, pp. 44–54, 2014.

[3] M. Gellert, K. I. Gries, J. Sann, et al., "Impedance spectroscopic study of the charge transfer resistance at the interface between a LiNi0.5Mn1.5O4 high-voltage cathode film and a LiNbO3 coating film," Solid State Ionics, vol. 287, pp. 8–12, 2016.

[4] A. A. Piarristeguy, M. Ramonda, and A. Pradel, "Local electrical characterization of Ag conducting chalcogenide glasses using electric force microscopy," Journal of Non-Crystalline Solids, vol. 356, pp. 2402–2405, 2010.

[5] A. J. Bhattacharyya, J. Fleig, Y. G. Guo, et al., "Local conductivity effects in polymer electrolytes," Advanced Materials, vol. 17, pp. 2630–2634, 2005.

[6] A. Layson, S. Gadad, D. Teeters, "Resistance measurements at the nanoscale: scanning probe ac impedance spectroscopy," Electrochimica Acta, vol. 48, pp. 2207–2213, 2003.

[7] A. R. Layson and D. Teeters, "Polymer electrolytes confined in nanopores: using water as a means to explore the interfacial impedance at the nanoscale," Solid State Ionics, vol. 175, pp. 773–780, 2004.

[8] J. Fleig, S. Rodewald, and J. Maier, "Spatially resolved measurements of highly conductive and highly resistive grain boundaries using microcontact impedance spectroscopy," Solid State Ionics, vol. 136–137, pp. 905–911, 2000.

[9] L. E. Walther, N. E. Israeloff, E. V. Russel, et al., "Mesoscopic-scale dielectric relaxation at the glass transition," Physical Review B, vol. 57, pp. R15112–R15115, 1998.

[10] A. Schirmeisen, A. Taskiran, H. Fuchs, et al., "Probing ion transport at the nanoscale: time-domain electrostatic force spectroscopy on glassy electrolytes," Applied Physics Letter, vol. 85, pp. 2053–2055, 2004.

[11] R. O'Hayre, G. Feng, W. D. Nix, et al., "Quantitative impedance measurements using atomic force microscopy," Journal of Applied Physics, vol. 96, pp. 3540–3549, 2004.

[12] J. Fleig and J. Maier, "Local conductivity measurements on AgCl surfaces using microelectrodes," Solid State Ionics, vol. 85, pp. 9–15, 1996.

[13] R. O'Hayre, M. Lee, and F. B. Prinz, "Ionic and electronic impedance imaging using atomic force microscopy," Journal of Applied Physics, vol. 95, pp. 8382–8392, 2004.

[14] B. Roling and S. Murugavel, "Bulk and interfacial ionic conduction in a LiAlSiO4 glass ceramic containing nano- and microcrystallites," Zeitschrift fuer Physikalische Chemie, vol. 219, pp. 23–33, 2005.

[15] B. Roling, A. Schirmeisen, H. Bracht, et al., "Nanoscopic study of the ion dynamics in a LiAlSiO4 glass ceramic by means of electrostatic force spectroscopy," Physical Chemistry Chemical Physics, vol. 7, pp. 1472–1475, 2005.

[16] A. Schirmeisen, A. Taskiran, H. Fuchs, et al., "Interfacial ionic conduction in a nanostructured glass ceramic," Physical Review Letters, vol. 98, p. 225901, 2007.

[17] J. Krümpelmann, M. Balabajew, M. Gellert, et al., "Quantitative nanoscopic impedance measurements on silver-ion conducting glasses using atomic force microscopy combined with impedance spectroscopy," Solid State Ionics, vol. 198, pp. 16–21, 2011.

[18] J. Krümpelmann, H. Reinhard, C. Yada, et al., "AFM tip-induced metal particle formation on laser-structured and on unstructured surfaces of solid-state ion conductors," Solid State Ionics, vol. 234, pp. 46–50, 2013.

Abdulkadir Sanli, Abderrahmane Benchirouf, Christian Müller and Olfa Kanoun

AC Impedance Investigation of weg Multi-walled Carbon Nanotubes/PEDOT:PSS Nanocomposites Fabricated with Different Sonication Times

Abstract: In this study, the effect of the sonication time on the electrical properties of multi-walled carbon nanotubes (MWCNTs) combined with poly(3,4-ethylenedioxythiophene)/poly(4-styrene-sulphonate (PEDOT:PSS) was investigated by impedance spectroscopy in the frequency range from 40 Hz to 110 MHz. The MWCNTs/PEDOT:PSS containing 0.0125, 0.025, 0.05, 0.075 and 0.1 wt.% were prepared with different sonication times 15, 30, 45 and 60 min and the nanocomposites were deposited on a flexible substrate (Kapton HN) by the drop casting technique. The real part of the impedance responses shows that the nanocomposites for 30 min sonication time have better electrical characteristics and higher conductivity. In addition, the dispersion quality of the nanocomposites as a function of sonication time was investigated by scanning electron microscopy (SEM). The SEM images proved that the homogeneity of the nanocomposites depends not only on the MWCNTs concentration but also on the sonication time. Consequently, a corresponding equivalent circuit was proposed on the basis of experimental results aiming to reveal the physical behaviour of the nanocomposite.

Keywords: Multi-walled carbon nanotubes, sonication time, PEDOT:PSS, impedance spectroscopy, equivalent circuit, modeling

1 Introduction

Intrinsic conductive polymers (ICPs) such as poly(3,4-ethylenedioxythiophene) and poly(4-styrene-sulphonate) known as PEDOT:PSS have drawn a great attention owing to their versatile applications in many engineering applications such as transparent electrodes [1], solar cells [2–4], supercapacitors [5, 6] and strain sensors [7, 8]. By taking the advantages of PEDOT:PSS, it is possible to enhance the processability and physical characteristics of the nanocomposite when it is combined with nanofillers such as carbon black, graphite, metal particles and carbon nanotubes (CNTs). Among above nanofillers, CNTs are considered a prominent candidate because of their excellent physical and chemical properties. Considerable features of CNTs, that is high interfacial attraction with polymer matrix and high aspect ratio enable the formation of conductive paths at very low nanofiller concentration. However, CNTs have poor solubility in aqueous solutions as well as they have a tendency to agglomerate because of the strong van der Walls interaction between the nanotubes, which declines their electrical and mechanical properties [9, 10]. Thus, controlling the processing parameters such as filler concentration and sonication time for functionalizing the carbon nanotubes is an important step to obtain a homogeneous CNT-based nanocomposite.

In the past few years, a number of studies concerning the processing parameters of CNT-based nanocomposites have been conducted. Bu et al. [11] have investigated the effect of sonication time, filler concentration and annealing process on electrical properties of carbon nanotubes, and it was found that sonication time plays a crucial role in the homogeneity and quality of the CNT-based films. Montazeri et al. [12] have studied the influence of sonication time on the mechanical properties of MWCNT-based epoxy nanocomposites for strain-sensing applications, and it was pointed out that sonication time has a considerable influence on tensile strength and fracture toughness of the nanocomposite. Suave et al. [13] observed that for carboxylic-functionalized single-walled carbon nanotubes (c-SWCNTs), increasing the sonication time from 20 to 40 min enhances the tensile strengths from 20.5 ± 1.4 GPa to 31.5 ± 2.2 GPa. In our previous study [8], for MWCNTs/PEDOT:PSS nanocomposites the highest conductivity and reproducibility were achieved at 30 min sonication time by direct mixing of the MWCNTs with PEDOT:PSS. However, there are currently limited works addressing the investigation of the sonication time on the dispersion quality of nanocomposite in the frequency domain.

In this chapter, the goals are to investigate the effects of the sonication time on the dispersion quality and to understand the complex conduction mechanism of the MWCNTs/PEDOT:PSS nanocomposite by using impedance spectroscopy.

2 Experimental

2.1 Materials and Nanocomposite Preparation

The PEDOT:PSS was purchased from Heraus (Leverkusen, Germany). The MWCNTs were obtained from Southwest Nanotechnology with 95% purity and 6–9 nm outer diameter and less than 1 μm length and they were used without any chemical treatment.

Fabrication process started with mixing 1.3 wt.% of PEDOT:PSS directly with MWCNTs with different concentrations (0.0125, 0.025, 0.05, 0.75 and 0.1 wt.%). Then, the mixtures are sonicated for 15, 30, 45 and 60 min by using a horn sonicator (Bandelin GM 3200). A sonication power of 30 W was applied with 50% duty cycle at room temperature. After the dispersion process, the MWCNTs/PEDOT:PSS suspensions were centrifuged at 5,000 rpm for 40 min using a Sigma 2-16 PK centrifuge aiming to disperse bundled, unbundled MWCNTs and remove the remaining impurities. The fabrication process for the MWCNTs/PEDOT:PSS dispersion is illustrated in Fig. 1.

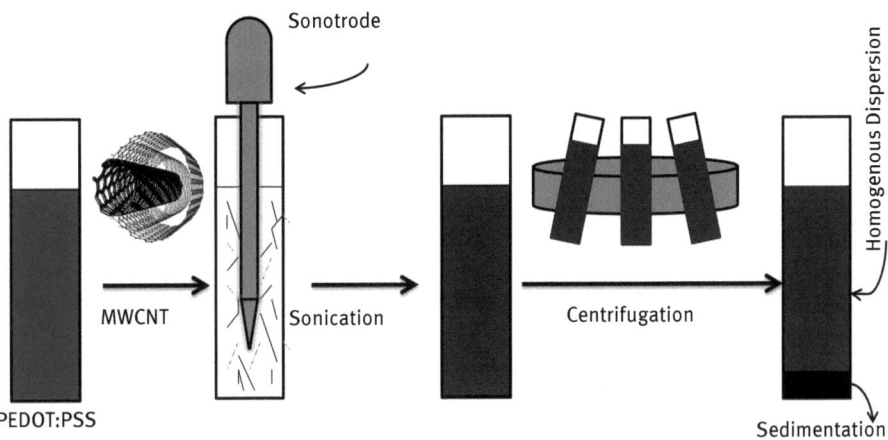

Fig. 1. Schematic illustration of the fabrication steps for MWCNTs/PEDOT:PSS nanocomposites.

Prior to the deposition, a thin flexible polyimide substrate (Kapton HN) was cleaned with ethanol followed by distilled (DI) water and finally dried with Nitrogen (N_2). After the substrate treatment, the thin substrate was covered with isolation foil and a pre-defined pattern was cut into slice of 2 cm × 0.5 cm. An optimized amount of dispersion (62.5 µl) was deposited on the substrate by means of an electronic pipette (Brand GmbH, Wertheim, Germany). The thin films were then dried overnight in a clean chamber. After the deposition and drying process, the thin isolation foils were removed and silver paste was applied on both sides of the film to reduce contact resistance. The thin film fabrication steps are shown in Fig. 2.

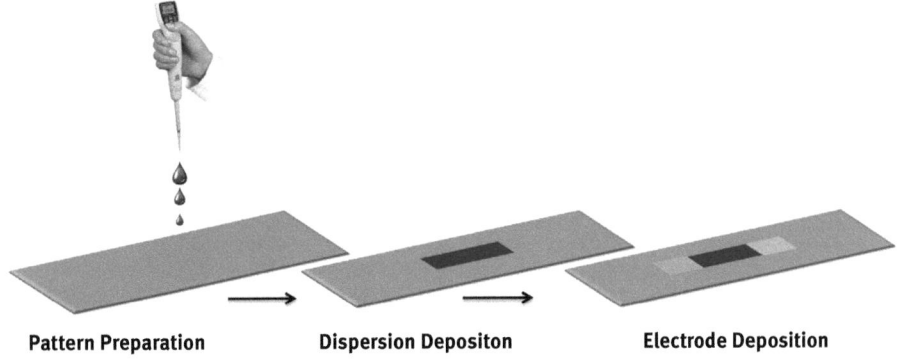

Fig. 2. The steps of thin film fabrication and electrode deposition.

2.2 Measurement Set-up

The impedance measurements of thin films were carried out at room temperature using Agilent 4294A impedance analyzer in the frequency range of 40 Hz to 110 MHz. The thin films were connected to the impedance analyzer device. For the impedance measurements the AC amplitude and the DC-bias were set to 10 mV and 0 V, respectively.

3 Results and Discussions

3.1 Morphological Analysis

The morphological investigations of MWCNTs/PEDOT:PSS nanocomposites were examined by scanning electron microscopy (SEM). The SEM images of nanocomposites for 0.1 wt.% of filler concentration at different sonication times are shown in Fig. 3.

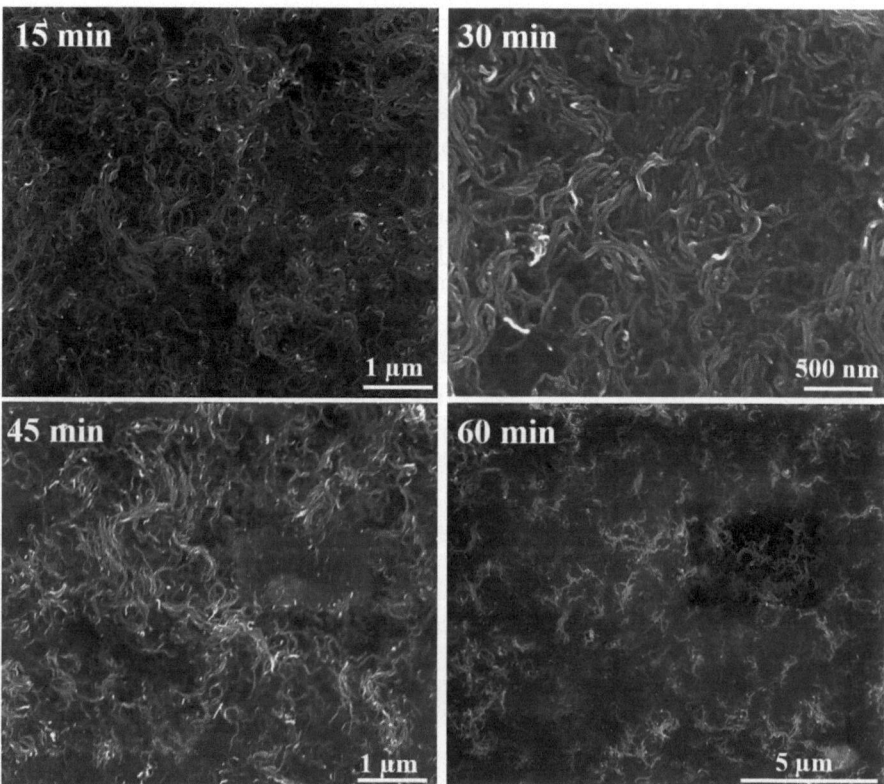

Fig. 3. SEM images of MWCNTs/PEDOT:PSS nanocomposites with 0.1 wt.% MWCNTs at different sonication times.

In case of the nanocomposites fabricated at 30 min sonication time, the MWCNTs are fairly visible in comparison with the other sonication times and the MWCNTs are well dispersed through the PEDOT:PSS polymer. These results were compared with the electrical properties investigated by impedance spectroscopy in the following section.

3.2 Electrical Characteristics

As it is mentioned above, CNTs have a tendency to bundle because of van der Walls interaction between the tubes. The formation of bundles causes poor electrical and mechanical properties of CNTs within a polymer matrix. Related to the fabrication of the dispersion by sonication, a sound wave that propagates the suspensions is applied to overcome the van der Walls forces. Hence, the MWCNTs will directly attach to PEDOT molecules since PEDOT is also not soluble in aqueous solution. However, high sonication power or time may harm the CNTs and introduce many defects such as breaking of the CNTs [14]. Therefore, the dispersion process should be well understood and optimized to obtain good electrical characteristics from the nanocomposites.

From the Nyquist plots shown in Figs 4 and 5, it can be seen that the nanocomposites at 30 min sonication time show the lowest resistance, resulting in the highest conductivity for the 0.05 wt.% and 0.1 wt.% filler contents. It is important to note that, above 30 min sonication time, the MWCNTs in the suspensions might be spreading more from each other, resulting in a higher tunneling resistance, or more defects might be introduced. For sonication times less than 30 min, most of the CNTs remain unbundled and therefore the measured resistance values are higher.

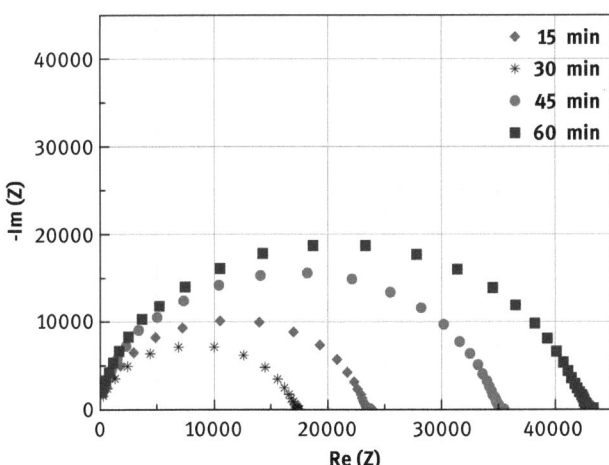

Fig. 4. Nyquist plot of the MWCNTs/PEDOT:PSS with 0.05 wt.% MWCNTs at different sonication times.

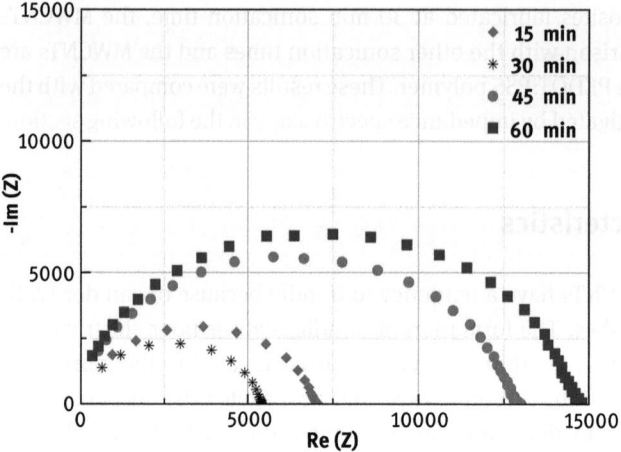

Fig. 5. Nyquist plot of the MWCNTs/PEDOT:PSS with 0.1 wt.% MWCNTs at different sonication times.

3.3 Equivalent Circuit Modelling

There are numerous experimental and theoretical studies mainly focusing on DC domain concerning the investigation of electrical properties of randomly distributed CNT networks [15–18]. The variation of the electrical properties under CNT loading and processing parameters, for example sonication time can be estimated through an appropriate resistance and capacitance network. In order to understand the complex transport mechanism of the MWCNTs/PEDOT:PSS nanocomposite, a 3D physical model was employed to estimate the function of nanofillers and polymer at low and high loadings individually.

At low CNTs loading, that is below 0.0125 wt.%, the gap between CNTs and CNT aggregates is very large and thus the tunneling resistance is high. As a result, the resistivity of the nanocomposite is approximately equal with PEDOT:PSS resistivity. As the CNT content increases, the gaps between CNTs decrease and the percolation [19] takes place where the conductive paths are formed, and resistivity of nanocomposite decreases sharply. However, because of the PEDOT:PSS polymer around the CNTs, the electrons passage between the CNTs is due to the tunneling. In CNT-based nanocomposites, the tunneling resistance is found to be dominant in electrical conductivity of nanocomposites when the inter-tube distance is not larger than 2 nm [20]. The tunneling resistance can be expressed according to the Simmon's theory [21] by:

$$R_{tunnel} = \frac{V}{AJ} = \frac{h^2 d}{A_{CNT} e^2 \sqrt{2m\lambda}} e^{\frac{4\pi d}{h}\sqrt{2m\lambda}} \qquad (1)$$

where J is the tunneling density, V is the electrical potential difference, d is the distance between CNTs, e is the electron charge, h is the Planck's constant, m is the electron mass, λ is the height of the barrier and A_{CNT} is the cross-sectional area of the tunnel. On the other hand, the effect of the resistance of CNTs (R_{CNT}) and agglomerates is found to be quite low compared with the tunneling resistance (R_{tunnel}) and capacitance [22]. Therefore, the contribution of R_{CNT} to the entire nanocomposite resistivity is omitted from the equivalent circuit.

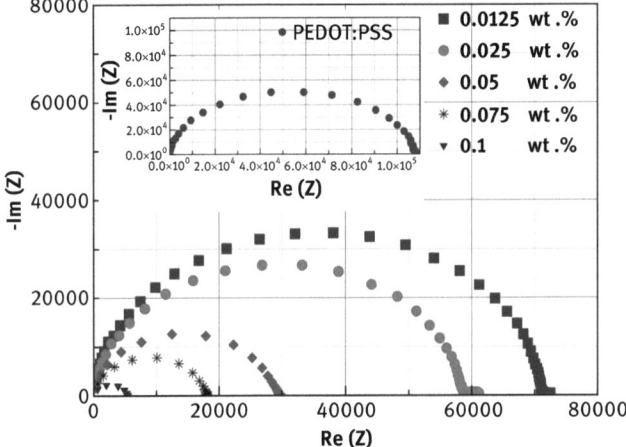

Fig. 6. Nyquist plot of (a) PEDOT:PSS and (b) MWCNTs/PEDOT:PSS at 30 min sonication time and different MWCNTs concentrations.

The overall impedance responses of the MWCNTs/PEDOT:PSS nanocomposites with 30 min sonication time and different CNT concentrations are shown in Fig. 6 and the corresponding equivalent circuit model for MWCNTs /PEDOT:PSS nanocomposites is illustrated schematically in Fig. 7.

From the equivalent circuit (see Fig. 7B), the total impedance of the equivalent circuit model can be written as:

$$Z_{total} = \frac{1}{1/R_{tunnel} + Q(j2\pi f)^\alpha} \qquad (2)$$

where the term $1/Q(j2\pi f)^\alpha$ corresponds to the constant phase element (CPE) that is used mostly in modelling of heterogeneous systems in the frequency domain response. The CPE parameter becomes a capacitance when $\alpha=1$ [23]. The fitted impedance response for 0.1 wt.% MWCNTs/PEDOT:PSS sonicated for 30 min is shown in Fig. 8 and the corresponding fitting parameters are summarized in Tab. 1.

From the fitting results and corresponding errors, it can be clearly observed that the proposed equivalent circuit is in a good agreement with the experimental data.

The obtained fitted parameters R_{tunnel} and Q parameter α are plotted in Fig. 9. It can be clearly stated that the R_{tunnel} value decreases with the increase of MWCNTs

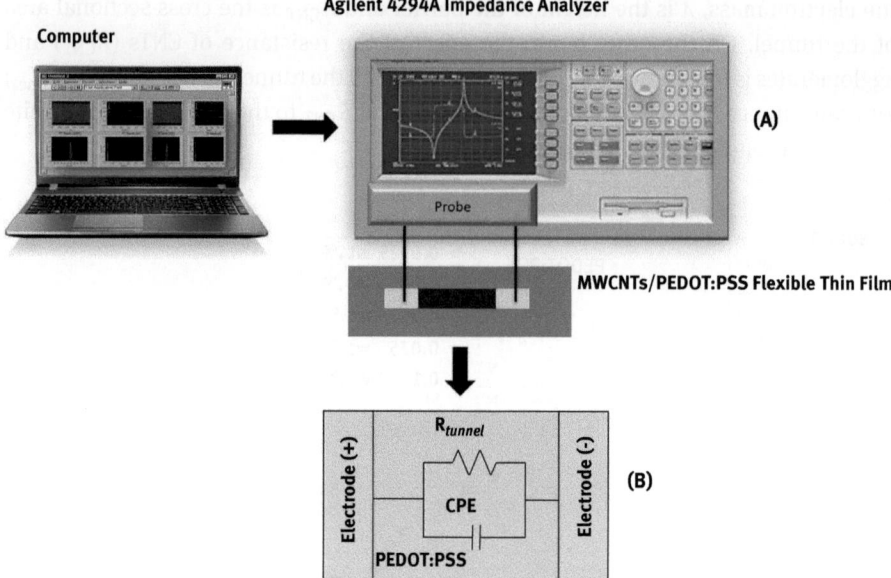

Fig. 7. NMWCNTs/PEDOT:PSS thin film physical model. (A) Schematic view of the measurement set-up. (B) Corresponding equivalent circuit model.

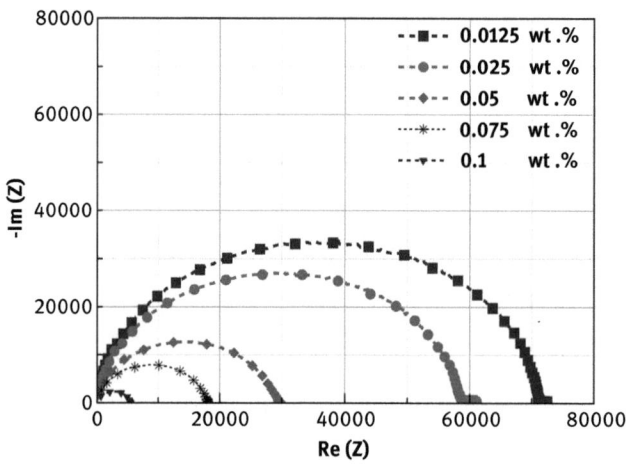

Fig. 8. Equivalent circuit fitting for the MWCNTs/PEDOT:PSS nanocomposite at different concentrations and 30 min sonication time.

Tab. 1. Fitting parameters for MWCNTs/PEDOT:PSS nanocomposites sonicated at 30 min for different loadings

Concentration (wt.%)	R_{tunnel}	α	Error (%)
0	108.08	0.97	0.01
0.0125	71.56	0.96	0.01
0.025	59.21	0.95	0.01
0.05	29.46	0.92	0.02
0.075	18.11	0.92	0.01
0.1	5.42	0.90	0.01

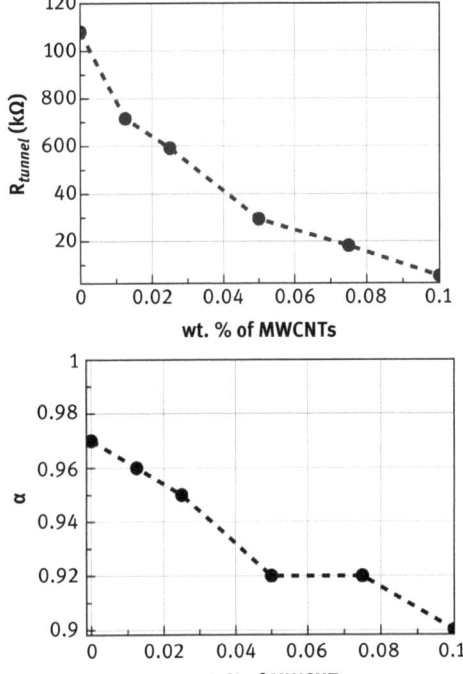

Fig. 9. Resistance change of MWCNTs/PEDOT:PSS nanocomposite as a function of MWCNTs concentration at 30 min sonication time (left). CPE parameter α as a function of MWCNTs content (right).

concentration, which is attributed to the formation of more conductive paths within the nanocomposite. Furthermore, the decrease in CPE parameter α may be attributed to the increase of inhomogeneity of the nanocomposite with MWCNTs loading.

4 Conclusions and Outlook

The MWCNTs/PEDOT:PSS nanocomposites with nanotube loading up to 0.1 wt.% were prepared and deposited on flexible thin substrate using drop casting technique. The effects of sonication time on the dispersion quality and electrical properties of nanocomposite were examined by SEM and impedance spectroscopy. The SEM images for different sonication times revealed that nanocomposite prepared at 30 min and 30 W power shows better homogeneity, which was also confirmed by the Nyquist plot of the impedance data. The real part of the impedance plot proved also that the nanocomposites prepared at 30 min sonication time show better electrical properties and conductivity. It was confirmed that the proposed equivalent circuit has a good agreement with the experimental data and it was appropriate to estimate the complex impedance response of these nanocomposites. Further studies with impedance spectroscopy will be concentrated on the adaptation of physical model under tensile strain, including the environmental effects, for example temperature and humidity.

Acknowledegment: The authors gratefully acknowledge Bilgehan Demirkale for designing the measurement set-up for AC measurements of the thin films, Paul Büschel and Thomas Günther for their technical support and fruitful discussions.

5 References

[1] M. Vosgueritchian, D. J. Lipomi, and Z. Bao, "Highly conductive and transparent PEDOT: PSS films with a fluorosurfactant for stretchable and flexible transparent electrodes," Advanced Functional Materials, vol. 22, no. 2, pp. 421–428, 2012.

[2] W. Hong, Y. Xu, G. Lu, et al., "Transparent graphene/PEDOT-PSS composite films as counter electrodes of dye-sensitized solar cells," Electrochemistry Communications, vol. 10, no. 10, pp. 1555–1558, 2008.

[3] Y. H. Kim, C. Sachse, M. L. Machala, et al., "Highly conductive PEDOT: PSS electrode with optimized solvent and thermal post-treatment for ITO-free organic solar cells," Advanced Functional Materials, vol. 21, no. 6, pp. 1076–1081, 2011.

[4] S. H. Eom, S. Senthilarasu, P. Uthirakumar, et al., "Polymer solar cells based on inkjet-printed PEDOT: PSS layer," Organic Electronics, vol. 10, no. 3, pp. 536–542, 2009.

[5] D. Antiohos, G. Folkes, P. Sherrell, et al., "Compositional effects of PEDOT-PSS/single walled carbon nanotube films on supercapacitor device performance," Journal of Materials Chemistry, vol. 21, no. 40, pp. 15987–15994, 2011.

[6] A. K. Cuentas Gallegos and M. E. Rincon, "Carbon nanofiber and PEDOT-PSS bilayer systems as electrodes for symmetric and asymmetric electrochemical capacitor cells," Journal of Power Sources, vol. 162, no. 1, pp. 743–747, 2006.

[7] G. Latessa, F. Brunetti, A. Reale, et al., "Piezoresistive behaviour of flexible PEDOT: PSS based sensors," Sensors and Actuators B: Chemical, vol. 139, no. 2, pp. 304–309, 2009.

[8] A. Benchirouf, A. Sanli, I. El-Houdaigui, et al., "Evaluation of the piezoresistive behavior of multifunctional nanocomposites thin films," in 11th International MultiConference on Systems, Signals Devices (SSD), IEEE, Castelldefels-Barcelona, Spain, 2014, pp. 1–4.

[9] L. F. C. Pereira and M. S. Ferreira, "Electronic transport on carbon nanotube networks: a multiscale computational approach," Nano Communication Networks, vol. 2, no. 1, pp. 25–38, 2011.

[10] L. A. Girifalco, M. Hodak, and R. S. Lee, "Carbon nanotubes, buckyballs, ropes, and a universal graphitic potential," Physical Review B, vol. 62, no. 19, p. 13104, 2000.

[11] L. Bu, J. Steitz, N. Dinh-Trong, et al., "Influence of processing parameters on electrical properties of carbon nanotube films," in Proceedings of the 7th International Multi-Conference on Systems, Signals and Devices–Sensors, Circuits & Instrumentation Systems, Amman, Jordanien, 27–30 June 2010.

[12] J. J. A. Montazeri, A. Khavandi, and A. Tcharkhtchi, "An investigation on the effect of sonication time and dispersion medium on the mechanical properties of MWCNT/epoxy nanocomposites," Advanced Materials Research, vol. 264–265, no. 9, pp. 1954–1959, 2011.

[13] J. Suave, L. A. F. Coelho, S. C. Amico, et al., "Effect of sonication on thermo-mechanical properties of epoxy nanocomposites with carboxylated-SWNT," Materials Science and Engineering: A, vol. 509, no. 1, pp. 57–62, 2009.

[14] O. Kanoun, C. Müller, A. Benchirouf, et al., "Flexible carbon nanotube films for high performance strain sensors," Sensors, vol. 14, no. 6, pp. 10042–10071, 2014.

[15] D. C. Lee, G. Kwon, H. Kim, et al., "Three-dimensional Monte Carlo simulation of the electrical conductivity of carbon nanotube/polymer composites," Applied Physics Express, vol. 5, no. 4, p. 45101, 2012.

[16] S. I. White, B. A. DiDonna, M. Mu, et al., "Simulations and electrical conductivity of percolated networks of finite rods with various degrees of axial alignment," Physical Review B, vol. 79, no. 2, p. 24301, 2009.

[17] F. Dalmas, R. Dendievel, L. Chazeau, et al., "Carbon nanotube-filled polymer composites. Numerical simulation of electrical conductivity in three-dimensional entangled fibrous networks," Acta materialia, vol. 54, no. 11, pp. 2923–2931, 2006.

[18] Y. Yu, G. Song, and L. Sun, "Determinant role of tunneling resistance in electrical conductivity of polymer composites reinforced by well dispersed carbon nanotubes," Journal of Applied Physics, vol. 108, no. 8, p. 84319, 2010.

[19] G. Grimmett, What is Percolation? Berlin: Springer, 1999.

[20] C. Li, E. T. Thostenson, and T.-W. Chou, "Dominant role of tunneling resistance in the electrical conductivity of carbon nanotube-based composites," Applied Physics Letters, vol. 91, no. 22, p. 223114, 2007.

[21] J. G. Simmons, "Generalized formula for the electric tunnel effect between similar electrodes separated by a thin insulating film," Journal of Applied Physics, vol. 34, no. 6, pp. 1793–1803, 1963.

[22] B. De Vivo, P. Lamberti, G. Spinelli, et al., "Numerical investigation on the influence factors of the electrical properties of carbon nanotubes-filled composites," Journal of Applied Physics, vol. 113, no. 24, p. 244301, 2013.

[23] M. E. Orazem, I. Frateur, B. Tribollet, et al., "Dielectric properties of materials showing constant-phase-element (CPE) impedance response," Journal of The Electrochemical Society, vol. 160, no. 6, pp. C215–C225, 2013.

Part IV: **Bioimpedance**

Marco Carminati
From Counting Single Biological Cells to Recovering Photons: The Versatility of Contactless Impedance Sensing

Abstract: Two modern applications of contactless impedance sensing at the microscale are reviewed. In particular, the cross-disciplinary evolution of contactless conductivity sensing between planar micro-electrodes from cell biology to a novel application in the field of solid-state silicon photonics is presented. The versatility of the same equivalent model, identified by means of impedance spectroscopy, is here highlighted. The presence of single cells in ionic solutions can be detected either in static (for monitoring the growth of a colony of adherent cells) or dynamic conditions (in micro-fluidic impedance flow cytometry), thanks to the contrast in conductivity between the insulating cell volume and the physiological solution, probed bypassing the electrochemical double-layer interfacial capacitance. Analogously, it is possible to leverage impedance detection in order to monitor the power of light propagating in silicon waveguides and weakly interacting with the wall interfaces. Here, as well, it is necessary to bypass the insulating thin silica cladding by means of a couple of planar micro-electrodes in order to probe the conductivity of the silicon core (in the nS range), increased by free carriers (from 1 to 10^3 per μm of waveguide) generated through photon absorption mediated by intra-gap energy states, which are created by the defects at the semiconductor surface. Such a non-invasive power monitor, featuring a detection limit of −35 dBm, 80 dB of dynamic range, sub-ms response speed and requiring no process modification, can be easily parallelized and allows for closed-loop control of optical devices. In spite of the apparent distance between these two examples, common design criteria for both the micro-electrodes geometry and the sensing electronics are here briefly discussed.

Keywords: Impedimetric sensors, impedance flow cytometry, silicon photonics

1 Introduction

Contactless impedance sensing represents a well-consolidated approach among industrial sensors. For example, capacitive level gauging in tanks leverages the difference in dielectric constant between air ($\epsilon_r = 1$) and the liquid filling the space between the capacitor electrodes. The size of the electrodes is thus adapted to the tank size, and their shape can be arranged in several ways: parallel plates, coaxial

---

Marco Carminati, Dipartimento di Elettronica, Informazione e Bioingegneria, Politecnico di Milano, Milano, Italy

cylinders, coplanar bands, etc. Beyond simplicity and linearity, this solution offers the major advantage that the sensing electrodes can be isolated from the liquid, for instance, placed on the outer surface of the tank, which consequently must not be made of metal in this case (Fig. 1A). Electrode fouling, corrosion and any other aging or degradation effect are thus completely avoided. In addition to detecting the presence (i.e., the volume) of the liquid, impedance can be also employed to detect the properties of the liquid, either its conductance or permittivity, for instance, in multi-phase solutions (such as emulsions or mixtures of immiscible fluids, liquids with gas bubbles or lubricant oils containing solid debris or water droplets, etc.).

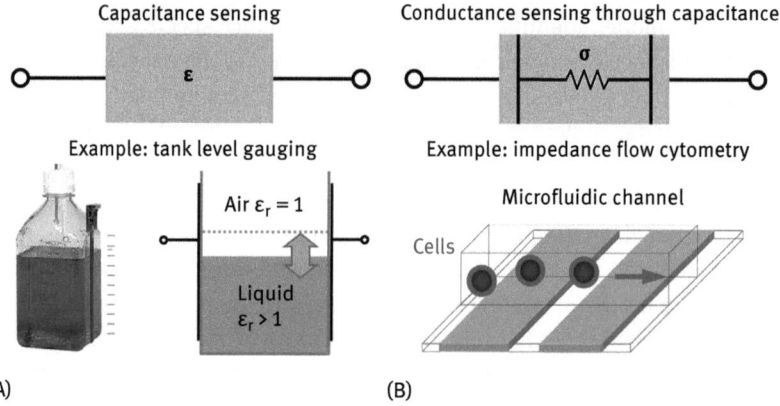

Fig. 1. Comparison of contactless impedance sensing approaches: (A) classic dielectric sensing versus (B) novel AC-coupled resistive sensing.

By extending a single bipolar measurement to a set of properly arranged multiple electrodes, for instance, around a pipe, it is possible to perform impedance tomography for contactless imaging the inner volume of the system under investigation, such as a pipe section or the human body. Several sensing approaches can be adopted. In this chapter, we focus on a particular case of contactless sensing, demonstrating the versatility, as well as the most significant design rules and challenges, of the same equivalent model in two apparently very different applications of impedance detection at the micro-scale: from cell biology to solid-state integrated photonics.

2 A General Equivalent Impedance Model

An important distinction between two different contactless sensing approaches must be initially highlighted. As illustrated in Fig. 2, two different situations can be encountered. In the first case (a), the measurand quantity is the capacitance of the sensor (i.e., the dielectric constant ε_r of the material between the electrodes), and the equivalent model is simply a capacitor C_M. In the second case (b), instead, the quantity of interest is the conductivity σ of the material, which is accessed in a contactless way, i.e., through a capacitor in series. Usually this capacitor is not a physical component but represents the capacitive coupling between the electrode and the bulk material. Consequently, the equivalent model is more complicated and includes the access capacitance C_A in series to the resistance R_M of the material. The bulk material also has a dielectric behavior and, thus, another capacitor should be added in parallel to the resistance, but it can be neglected since, in this context, the material generally displays a conductive rather than dielectric behavior. Furthermore, another capacitance C_S is always present between the electrodes, usually because of the direct parasitic coupling between the electrodes themselves and between the external connections. In standard setups, with cables and metal tracks on printed circuit boards, C_S is in the pF range. It can be significantly reduced (down in the fF range) only if the front end of the detection circuit is miniaturized (i.e., integrated on a single micro-electronic chip) in order to be placed extremely close to the electrodes, which are connected to the electronic chip via ultra-short bonding wires [1, 2].

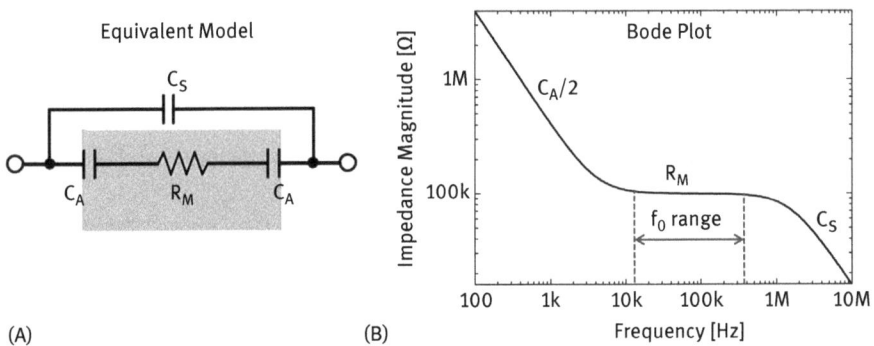

Fig. 2. (A) Equivalent model of the configuration for measuring the resistance R_M through the access capacitance C_A at the frequency f_0 to be properly chosen in the plateau region visible in the Bode Plot (B).

Here we focus on the second contactless approach, which finds primary application in bio-impedance sensing. From the analysis of impedance magnitude (shown in the Bode plot of Fig. 2B), it can be clearly observed that the value of the resistance can be accurately probed only within a limited frequency range, i.e., the flat resistive plateau. In fact, because usually C_S is smaller than C_A, the optimal measuring frequency should be high enough to short C_A, but lower than the second pole. The presence of C_S poses an upper limit the resistive plateau because its shunts R_M at high frequency. Outside of this range, the presence of the capacitors cannot be neglected: the value of R_M can still be extracted by an accurate phase-sensitive detector such as a lock-in amplifier.

If the values of C_M and C_S are not known (for instance, because either the geometrical dimensions or the materials parameters of the system under investigation are unknown), it becomes extremely important to preliminary perform an impedance spectrum, in order to identify the equivalent model and thus choose the most adequate value of tracking frequency (f_0) to access the resistive plateau. Beyond offering insight into the sensor properties and allowing the identification of the equivalent impedance model, impedance spectroscopy is beneficial also for the design of the electronic instrumentation. In fact, when employing a current reader to amplify the current flowing in the impedance (at the sense electrode) upon application (to the force electrode) of a voltage stimulation (generally a single sinusoidal tone or more articulated AC waveforms), its performances such as noise (i.e., resolution), bandwidth (i.e., speed) and stability depend on the equivalent impedance connected at the current input node [3].

3 Impedance Sensing in Cell Biology

The high-pass model of Fig. 2 is massively used in electrochemistry and biosensing corresponding to the lumped equivalent model of the simplest form of electrode/electrolyte interface, i.e., the non-Faradaic one. In this case, C_A represents the double-layer capacitance C_{DL}, which models the dynamic alteration of the ionic distribution at the interface between an ideally polarizable electrode and a solution containing ions.

One particular family of measurements that leverage the contrast between the conductive ionic solutions and suspended particles is cell detection. In fact, at low frequency (below 1–10 MHz), biological cells can be considered as insulating objects, as long as the membrane capacitance ($\sim$0.01 pF/μm^2) is not shorted. Thus, in order to detect the variations of the solution conductance of the physiological buffer (for instance, PBS has a conductivity σ_{PBS} = 0.66S/m) due to the presence of cells interacting with the electric field between the electrodes, impedance must be measured at a sufficiently high frequency to shunt the double-layer capacitance.

This label-free and quantitative cell detection technique is implemented in two different contexts. The first concerns "static" cell culture systems (such as modern micro-fluidic versions of the classic Petri dishes) in which cells are cultured on the top of planar electrodes (Fig. 3A). If the cells belong to an adherent cell line, they attach to the electrode surface and spread on the top of it. Thus, by looking at the impedance of these electrodes, it is possible to monitor in real time the growth of the colony [4]. The progressive coverage of the electrode surface with cells produces an increase of impedance [5]. The full life cycle of the cells can be monitored, because after cell death, they detach from the surface, causing a decrease of impedance. This technique is thus widely adopted for label-free and automated screening of the responses of cells to drugs.

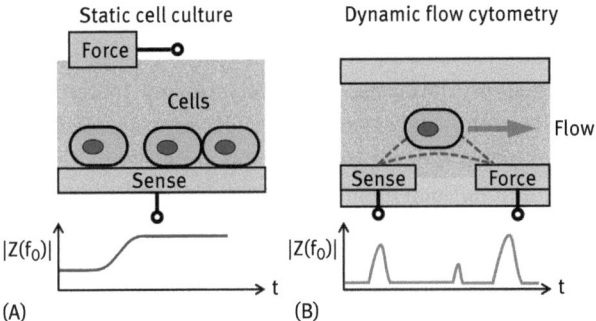

Fig. 3. Examples of application of impedance in cell detection by means of planar micro-electrodes: (A) static monitoring of the growth of a cell colony and (B) counting and sizing single cells focused in a stream focused inside a micro-fluidic channel.

Two electrodes configurations can be adopted: namely vertical or coplanar. In both cases, the optimal f_0 in terms of sensitivity in discriminating the presence of cells is at the corner between the capacitive and the resistive regimes, i.e., $f_0 = 1/(2\pi \cdot C_{DL} \cdot R_{SOL})$. We have theoretically and experimentally demonstrated that the coplanar arrangement is more sensitive, thanks to a stronger confinement of the electric field lines between the electrodes where adherent cells grow [6].

The second "dynamic" application scenario is that of impedance flow cytometry [7]. In this case, cells are suspended in the buffer and focused in a single stream, which flows in a micro-fluidic channel and interacts with a pair of electrodes (Fig. 3B). The passage of each cell produces a single pulse, whose amplitude is related to the volume of the cell. Thus, cell counting and sizing can be performed at high throughput (up to thousands cells per second) [8]. Here again, the micro-electrodes can be coplanar (patterned on the same chip) or arranged in a parallel-plate configuration. The latter provides uniform field lines, thus making the measurement independent of the position of the cell in the channel. However, it has two main drawbacks: (i) it requires accurate alignment between the two facing chips and (ii) it forces the channel to be extremely thin (so that the two plates are closer and the field is higher) and, thus, significantly more prone to clogging. On the contrary, the coplanar geometry is preferable for the simpler fabrication and for the possibility of making a larger channel, provided that some focusing mechanism is introduced to reduce the variability due to different distances from the electrodes.

4 Impedance-Based Light Monitoring

Taking inspiration from impedance flow cytometry, we devised the application of an analogous approach of micro-scale contactless conductivity sensing in the field of silicon photonics, belonging to the broader category of integrated optics. Integrated photonics encompasses, in fact, various emerging technologies for the manipulation of light on micro-fabricated solid-state platforms. A crucial problem in the case of silicon is the simple realization (from a fabrication point of view, especially in terms of compatibility with the standard CMOS process) of photodetectors and, in particular, of in-line light power monitors. Such monitors are necessary to monitor the operation of a photonic circuit and to provide a feedback signal that becomes pivotal in complex systems composed of several optical devices such as filters, resonators, interferometers, all needing independent control and stabilization. Because of fabrication tolerances, these devices need to be tuned to the desired working point before operation. Furthermore, being sensitive to environmental parameters (temperature in particular), they might need real-time closed-loop adjustment in order to reject slow drifts and fluctuations. Traditionally, in order to provide a monitor signal, a fraction of the light is tapped out of the device and directed to a photodetector that might be either integrated on chip or, more commonly for silicon, external, thus requiring an output fiber. Clearly this solution cannot be extended to hundreds or thousands of devices in cascade, because each monitor introduces additional loss.

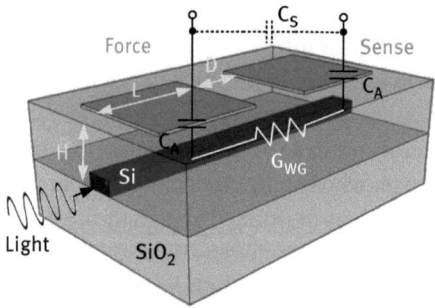

Fig. 4. CLIPP working principle: two micro-electrodes are fabricated on the top of the silica cladding to capacitively access (C_A) to the silicon waveguide conductance G_{WG} without perturbing light propagation.

Although silicon is transparent to mid-infrared radiation (i.e., the energy of photons is lower than the semiconductor energy gap), because of the presence of defects on the surface of the silicon crystal, at the interface between the Si core and the SiO_2 cladding, which introduces intra-gap energy states, some photons get inevitably absorbed. Consequently, the measurement of photogenerated carriers in a section of the waveguide provides a direct measure of the local intensity of light. However, a direct connection of a pair of metal or highly doped semiconductor electrodes to the silicon core produces a strong attenuation of the radiating field. In order to avoid any perturbation, we have recently proposed to adopt an innovative approach, i.e., to use contactless capacitive coupling to the waveguide [9]. As illustrated in Fig. 4, the novel sensor, named contactless integrated photonic probe (CLIPP), consists of a pair of planar electrodes placed on top of the silica cladding, at a vertical distance H (~1 μm) that provides a sufficient margin for attenuation of the evanescent field, leading to absolutely no perturbation. The electrodes of length L are separated by a distance D and are fabricated on the top of the cladding, i.e., on the same layer and with the same mask used to pattern metallic resistors employed as thermo-optic actuators in resonators or interferometers. Usually the thickness H is not a degree of freedom adjustable by the chip designer, because it is set by the constraints of the fabrication process. However, a first trade-off concerning H can be clearly identified: if H is too thin, the metal is too close to the waveguide, thus detrimentally interacting with the propagating field. On the other side, the larger H, the more difficult is the access to probe the waveguide electrical properties.

As anticipated, the equivalent circuit of the CLIPP is the same illustrated above. Each electrode has in series the vertical capacitance C_A through the silica cladding, giving access to the horizontal segment of the silicon waveguide comprised between the electrodes and characterized by its conductance G_{WG}, which is related to the light intensity. These parameters are actually distributed but, for simplicity, during the design of the device, they can be modeled with lumped components. In this case, i.e., with silicon-on-insulator technology, the total impedance is more complicated, because of the presence of another silica/silicon interface towards the conductive chip substrate, whose presence is always detrimental from the parasitic point of view and should be avoided when no active electronic circuits are monolithically integrated on the same chip [10]. The thickness of the bottom oxide is generally larger than H, so, at low frequency, this can be neglected. However, at high frequency, the current path through the substrate (either floating of grounded) becomes predominant (i.e., with a lower impedance) with respect to the path through the waveguide, making the measurement of G_{WG} more difficult, similarly to the effect of C_S.

Here as well, impedance spectroscopy is pivotal for the characterization of the device, the extraction of the values of the model parameters and, consequently, the choice of the optimal tracking frequency $f_0 > G_{WG}/(\pi \cdot C_A)$, which is in the MHz range, as illustrated in Fig. 5 where differential admittance spectra, measured for different input light power levels P, are reported. In order to get rid of the stray components (mainly of C_S and phase delays), the reference dark spectrum is subtracted from each measurement, and the values of ΔG_{WG} are taken in the flat regions. Fig. 5C shows the responses of three different CLIPP geometries, all described by a power law $\Delta G_{WG} = \alpha \cdot P^n$ with $n < 1$. CLIPPs fabricated on waveguides with the same cross section (width $w = 1$ μm) and different separation D (100 and 700 μm) show the same slope n in the bilogarithmic plot and ΔG_{WG} scales with $1/D$. Sensors with the same D but different waveguide widths ($w = 1$ μm and $w = 480$ nm) show similar values of ΔG_{WG}, but slightly different exponents n, which depend on the length of waveguide cross-sectional perimeter where defects are located.

This non-perturbative sensor, able to recover valuable information from photons that are unavoidably absorbed, has been successfully employed as a local power monitor, placed in various positions in the optical circuit normally not accessible by standard photodetectors [9] and also in closed-loop feedback systems for the stabilization of ring resonators [2].

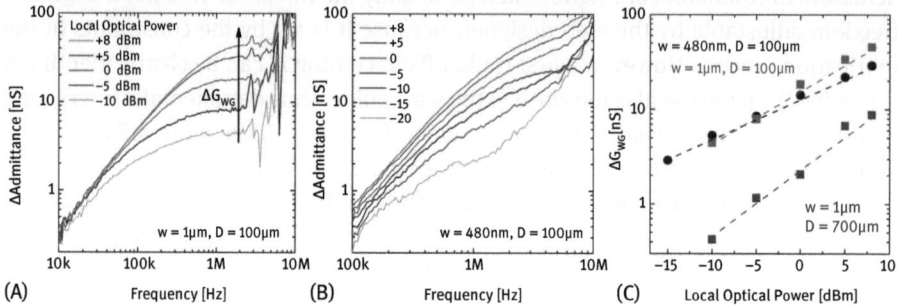

Fig. 5. Experimental characterization of the CLIPP static sensitivity by means of admittance spectroscopy (A) and (B), leading to the extraction of the sensor sub-linear response curve (C). The applied AC voltage is 2 V and the averaging time 1 s.

5 Common Design Criteria

In summary, as highlighted in the examples above, when designing a contactless impedance measuring system based on coplanar micro-scale electrodes, the following criteria must be followed:

- **Access capacitance:** in general, the access capacitance C_A should be made as large as possible, in order to extend the resistive plateau towards lower frequencies and allow a better access (i.e., with a lower series impedance) to the resistance that must be measured. In the case of the electrochemical double layer, because the specific capacitance (mostly dependent on the concentration of ions in solution and on the applied DC potential) is not a design parameter, C_A simply scales with the area of the electrodes. Neglecting surface treatments that increase the roughness of the electrode surface, larger electrodes provide a larger C_A, implying a larger area occupation, as well as larger parasitics; thus, an optimal value can be usually identified. Furthermore, in the case of the CLIPP sensor, the vertical distance H could represent an additional design parameter, provided that the fabrication process can be modified. In the latter case, in order to increase C_A, H should be reduced down to the minimum value that still avoids significant perturbation of the propagating optical field (500–1000 nm).

- **Sensing frequency:** the sensing frequency f_0 should be chosen to probe the resistive plateau if the latter has been identified by means of impedance spectroscopy. Thus, f_0 depends on the values of C_A and R_M. Considering that standard impedance sensing circuitry operates up to tens of MHz, this value should be taken as upper limit. Indeed, the GHz range is probed as well in the context of dielectric spectroscopy [11] by means of micro-wave circuits that are designed in a different way with respect to low-frequency ones. In particular, the main differences are in terms of component (high-quality passive components, higher operating bandwidths for active ones, often requiring single-transistor design, rather than operational amplifiers) board layout and connection considerations (whose length becomes comparable with the wavelength of the electrical signals).
- **Distance:** in the coplanar configuration, the separation D between the two closely spaced electrodes covered by a semi-infinite homogeneous material represents the most important design parameter because it determines the vertical extension of the fringing field. The vertical sensitivity thickness is approximately equal to D. Correspondingly, this vertical distance should be matched to the height of the passing objects, such as in the case of contactless impedance flow cytometry [8], or, in the case of adherent cells [5, 6] or dust particles depositing directly on the micro-electrodes [12], equal to the size of the particles themselves. When the distance between the electrodes is larger (as for the CLIPP), D sets the value of R_M. In the latter case, a shorter D implied higher current (i.e., higher G_{WG}) so apparently it should be minimized achieving a larger signal and a smaller area occupation. Unfortunately, a smaller D also implies a higher f_0, thus demanding for a balancing trade-off (usually in the 50–100 μm range for D, corresponding to the MHz range for f_0).
- **Modeling of parasitics and need for holistic miniaturization:** accurate and comprehensive modeling of all the impedance components, including the substrate material [10], is crucial. Wide-bandwidth impedance spectroscopy represents the best avenue to address this characterization, which should be performed in the initial phase of the development of any new impedimetric sensor. In particular, in the case of micro-scale transducers, when the size of the device scales down, its impedance accordingly increases, whereas the impedance of the connection lines typically stay constant, quickly becoming the dominant term and completely masking the transducer impedance and its variations. The best solution is the concurrent miniaturization of the detection electronics and of the interconnections, allowing a similar scaling of all the terms and, thus, a significant improvement of the achievable performance [1, 13, 14]. Analogously, shielding and separation between the forcing signal and the sensing line (current input) are crucial to reduce the impact of parallel stray paths, i.e., the value of C_S [15].
- **Electronics:** of course, in order to achieve the best signal-to-noise ratio, the signal should be maximized and the noise minimized. The former goal can be achieved

by applying the largest forcing signal sustainable by the device (for instance, about 100 mV for biochemical interface, up to several Volts for solid-state devices), where the limit is the breakdown field of the insulators surrounding the electrodes (where no DC current is flowing). One risk of using large forcing signals is related to the linearity of the system under investigation. If the system transfer function includes some kind of non-linearity, a large input stimulus will produce harmonics at the output, whose presence should be carefully checked in the frequency domain and avoided when tuning the value of the forcing amplitude. The impact of these harmonics on the measurement performance depends on the scheme adopted for the calculation of impedance: while the lock-in architecture (i.e., the synchronous demodulation locked with the forcing sinusoidal signal) is ideally robust with respect to higher harmonics, Fourier-based solutions are more prone to inaccuracies. Noise minimization requires accurate design of the current-sensing front-end [3], analysis and reduction of the impact of the noise of the forcing signal generator by means, for instance, of differential architectures [16] and accurate reduction of the total input capacitance, including the device capacitance and the connections stray capacitance [10, 17].

6 Conclusions

An new example of the versatility of contactless conductance sensing has been here reported. The major design issues and instrumentation aspects related to a recently proposed original light detection technique for solid-state photonic platforms have been briefly reviewed, showing the evolution of similar design guidelines from the context of micro-scale impedance sensing in cellular biology. Even if impedance tracking is performed at a single frequency, impedance spectroscopy should be preliminary performed to fit the proper equivalent model of the sensor, allowing an estimate of the setup parasitic terms, a proper low-noise design of the detection electronics and the choice of the optimal sensing frequency.

Very often, the non-specificity of impedance is considered as a limit, in particular within the context of impedimetric affinity biosensors where the impedance of an electrochemical interface is affected by the receptor response, as well as by many other non-specific interactions. Here, on the contrary, it has been shown that, thanks to the universality of impedance and its extremely wide range of applications, fruitful and inspiring cross-fertilization effects can be deployed, migrating similar design rationales and trade-offs across different fields.

Acknowledgments: I would like to thankfully acknowledge my mentors Prof. M. Sampietro and Prof. G. Ferrari and all the colleagues and co-workers in the different projects, both on impedimetric biological cell detection and on integrated photonics, in particular Prof. J. Voldman (MIT, USA), Prof. J. Emneus, Dr. A. Heiskanen, Dr. C. Caviglia (DTU, Denmark), Prof. A. Melloni and Dr. F. Morichetti (Politecnico di Milano, Italy). Financial support from Fondazione Rocca, Fondazione CARIPLO and from EU under projects EXCELL and BBOI is also acknowledged.

7 References

[1] M. Carminati, G. Ferrari, D. Bianchi, et al., "Femtoampere integrated current preamplifier for low noise and wide bandwidth electrochemistry with nanoelectrodes," Electrochimica Acta, vol. 112, pp. 950–956, 2012.

[2] S. Grillanda, M. Carminati, F. Morichetti, et al., "Noninvasive monitoring and control in silicon photonics using CMOS integrated electronics," Optica, vol. 1, pp. 129–135, 2014.

[3] M. Crescentini, M. Bennati, M. Carminati, et al., "Noise limits of CMOS current interfaces for biosensors: a review," IEEE Transactions on Biomedical Circuits and Systems, vol. 8, pp. 278–292, 2014.

[4] J. Wiegener, C. H. Keese, and I. Giaver, "ECIS as a non invasive means to monitor the kinetics of cell spreading to artificial surfaces," Experimental Cell Research, vol. 259, pp. 158–166, 2000.

[5] M. Vergani, M. Carminati, G. Ferrari, et al., "Multichannel bipotentiostat integrated with a microfluidic platform for electrochemical real-time monitoring of cell cultures," IEEE Transactions on Biomedical Circuits and Systems, vol. 6, pp. 498–507, 2012.

[6] M. Carminati, C. Caviglia, A. Heiskanen, et al., Theoretical and experimental comparison of microelectrode sensing configurations for impedimetric cell monitoring, ser. Lecture Notes on Impedance Spectroscopy, Vol. 4, Boca Raton, FL: CRC Press, 2013, pp. 75–82.

[7] T. Sun and H. Morgan, "Single-cell microfluidic impedance cytometry: a review," Microfluidics and Nanofluidics, vol. 8, pp. 423–443, 2010.

[8] M. Carminati, M. D. Vahey, A. Rottigni, et al., "Enhancement of a label-free dielectrophoretic cell sorter with an integrated impedance detection system," Proceedings of the 14th International Conference on Miniaturized Systems for Chemistry and Life Sciences (microTAS), pp. 1394–1396, 2010.

[9] F. Morichetti, S. Grillanda, M. Carminati, et al., "Non-invasive on-chip light observation by contactless waveguide conductivity monitoring," IEEE Journal of Selected Topics in Quantum Electronics, vol. 20, pp. 292–301, 2014.

[10] M. Carminati, M. Vergani, G. Ferrari, et al., "Accuracy and resolution limits in quartz and silicon substrates with microelectrodes for electrochemical biosensors," Sensors and Actuators B: Chemical, vol. 174, pp. 168–175, 2012.

[11] J. Chien and A. Niknejad, "Oscillator-based reactance sensors with injection-locking for high-throughput flow cytometry using microwave dielectric spectroscopy," IEEE Journal of Solid-State Circuits, vol. 51, pp. 457–472, 2016.

[12] M. Carminati, L. Pedalà, E. Bianchi, et al., "Capacitive detection of micrometric airborne particulate matter for solid-state personal air quality monitors," Sensors and Actuators A: Physical, vol. 219, pp. 80–87, 2014.

[13] D. Bianchi, M. Carminati, G. Ferrari, et al., "CMOS current amplifier for afm impedance sensing on chip with zeptofarad resolution," Proceedings of the 9th IEEE Ph.D. Research in Microelectronics and Electronics (PRIME), pp. 61–64, 2013.

[14] P. Ciccarella, M. Carminati, M. Sampietro, et al., "Multichannel 65 zF rms Resolution CMOS Monolithic Capacitive Sensor for Counting Single Micrometer-Sized Airborne Particles on Chip," IEEE Journal of Solid-State Circuits, in press.

[15] P. Ciccarella, M. Carminati, G. Ferrari, et al., "Impedance sensing CMOS chip for noninvasive light detection in integrated photonics," IEEE Transactions on Circuits and Systems II: Express Briefs, in press.

[16] M. Carminati, G. Gervasoni, M. Sampietro, et al., "Note: differential configurations for the mitigation of slow fluctuations limiting the resolution of digital lock-in amplifiers," Review of Scientific Instruments, vol. 87, pp. 026102, 2016.

[17] M. Carminati, G. Ferrari, A. P. Ivanov, et al., "Design and characterization of a current sensing platform for silicon-based nanopores with integrated tunneling nanoelectrodes," Analog Integrated Circuits and Signal Processing, vol. 77, pp. 333–343, 2013.

Hip Kõiv, Ksenija Pesti and Rauno Gordon
Electric Impedance Measurement of Tissue Phantom Materials for Development of Medical Diagnostic Systems

Abstract: In this chapter, we introduce a method to obtain tissue phantom materials with different electrical conductivities to have phantom organs with needed health state or pathologic condition. Those phantoms would be used in the development of medical diagnostic systems. Material developed in the research is gelatine-based and made of solidifying materials. Sodium chloride (NaCl) is used to adjust the electric properties of the phantoms while the resulting conductivity of the material is measured with Van der Pauw method in a custom box with four electrodes. Conductivity (σ) increased linearly with NaCl concentration, and we got reliable gelatine samples. The overall goal is to develop an easy recipe for gelatine phantoms that could be manipulated with NaCl and would be useful in the development of bioimpedance measurement systems.

Keywords: Van der Pauw method, gelatine phantom, electrical conductivity

1 Introduction

There are many phantom materials available for mimicking human tissues. They can be classified as hydrogels, organogels and flexible elastomer materials for radiological tissue parameters [1]. Agar, agarose and gelatine gels, manipulated by varying NaCl concentration from 0–1 mg/ml [4], are often used for mimicking tissue electrical parameters (for example, comparing research by Renn [9]). Agar and agarose both are derived from red algea, but agarose has undergone purification and does not have protein in it [9]. Gelatine is made of collagen in pork skins and bones. We decided to use gelatine, because it is easily available and cheap, and it melts at 35°C, when agar needs more heating (85°C). Similar research by Marchal et al. [10] suggests using gelatine phantoms for an easy and cheap simulation of most human tissues. The purpose is to have phantom organs available with desired electrical conductivities. This involves preparing the material and measuring electric impedance of the gelatine phantoms with Van der Pauw (vdP) method [2, 3]. Those phantoms would be used in the development of medical diagnostic systems, which use bioimpedance measurement with either electrodes or eddy currents to estimate the physiological state and processes in the patient. This would work just like

---

Hip Kõiv, Ksenija Pesti and Rauno Gordon, Thomas Johann Seebeck Department of Electronics, Tallinn University of Technology, Tallinn, Estonia

phantoms that are made for research in radiology, but with focus on electric and dielectric properties [10]. There are many bioimpedance-based diagnostic methods, and their number is growing because it is a location-dependent method – different aspects of bioimpedance measurement depend extensively on location on the patient anatomy, where the diagnosis is required. In addition, there are a lot of options in using different frequencies, different electrode shapes and configurations. Same applies to eddy current–based bioimpedance measurements. Therefore, there is a need for phantom materials with different electric properties in different shapes.

2 Materials and Methods

2.1 Gelatine Phantom Preparation

Gelatine phantom is used because of availability, simple production and the ability to manipulate with its electrical conductivity. To make the gelatine material, first, distilled water is added to the unflavoured gelatine powder for dilation. After that, it is heated in hot water bath (50°C) and stirred constantly till the solution of gelatine and water liquefies and becomes homogeneous. The rest of the ingredients (sodium chloride, ethyl 4-hydroxybenzoate, formaldehyde solution) are added, and the mixture is placed into the containers. Gelatine phantom is later placed in the refrigerator for storage. Gelatine gives mechanical strength to the phantom; salt and water control the conductivity and permittivity. We use a rotation machine during solidifying to obtain more homogeneous phantom. If we have reliable homogeneous phantoms with known conductivities, we can start to develop heterogeneous phantoms by adding different substances (for example, oil droplets [11]) or making layers of various materials. Tab. 1 shows an example of a phantom mixture used in the research.

Tab. 1. Phantom ingredients

Ingredient	Amount
Distilled water	250 ml for dilation
	333 ml added later
Sodium chloride (NaCl, 99+%)	A: 0 g
	B: 0.5 g
	C: 5 g
Ethyl 4-hydroxybenzoate (99+%, Sigma-Aldrich)	3 g
Formaldehyde solution (37 wt%, Sigma-Aldrich)	2 ml
Gelatine powder (240TM, Poland)	40 g

Ethyl 4-hydroxybenzoate is used as preservative, and formaldehyde solution is used to rise the melting temperature of the phantom. Different electrical conductivities of phantoms are achieved by adjusting different amounts of sodium chloride.

2.2 Plexiglas Enclosure with Electrodes

Enclosure was designed in AutoCad and Computer Numerical Control router milled from 10-mm Plexiglas. The dimensions of the inner cube are 50 mm × 50 mm × 50 mm to simplify the model when calculating phantom's electrical properties (Fig. 1).

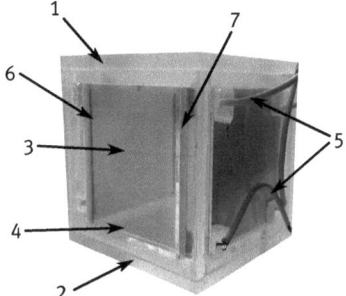

Fig. 1. Plexiglas cube for gelatine phantom. 1. Top side, 2. bottom side, 3. double-sided PCB, 4. hole to fill the cube with gelatine, 5. wired holes and pads, 6. electrodes for vdP measurements, 7. PCB inserting caps.

The double-sided printed circuit board (PCB) for vdP and eddy-current measurements was designed so that two long electrodes on the inside are connected to the pads on the outer side where wire leads are soldered. In addition, a planar coil was designed onto the PCB with corresponding pads on the outer side for connections. PCBs with different planar coils were made to test eddy-current measurements with different sensors (Fig. 2).

2.3 Van der Pauw Method

The vdP [2, 3] is used in this research to measure the electrical conductivity of designed gelatine samples. Four thin electrode stripes, on two specially designed PCBs on either side, have direct contact to the gelatine. When viewed top down as 2D, the point-like electrodes are exactly in the corner of the sample, which makes the vdP method compatible. Also, thickness of the sample must be constant and homogeneous [3]. vdP measurement is a very common resistivity method used in the semiconductor industry [4], and usually, vdP is used by other researchers on samples where the sample thickness is comparable to the surface dimensions. In our chapter, we are using given technique to measure the resistivity of gelatine cube. Research made by Kasl and Hoch [12] shows that the limiting sample thickness is about half the diameter of the samples when contacts are placed on the edge of the surface. When the contacts are located across the edges and the resistance of the electrodes is sufficiently low compared with the sample resistance, like in our research, vdP method is reliable [12].

vdP technique is based on the voltage and current ratio. Current I_{AB} is passed through the electrode A and extracted through B (Fig. 3(1)). To construct the needed resistance $R_{AB,DC}$ for vdP formula, measurement of the potential difference V_{DC}

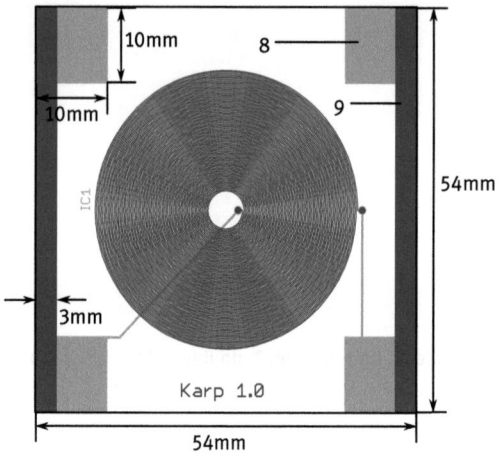

Fig. 2. Double-sided PCB design. 8. Soldering pad, 9. electrode on the bottom side.

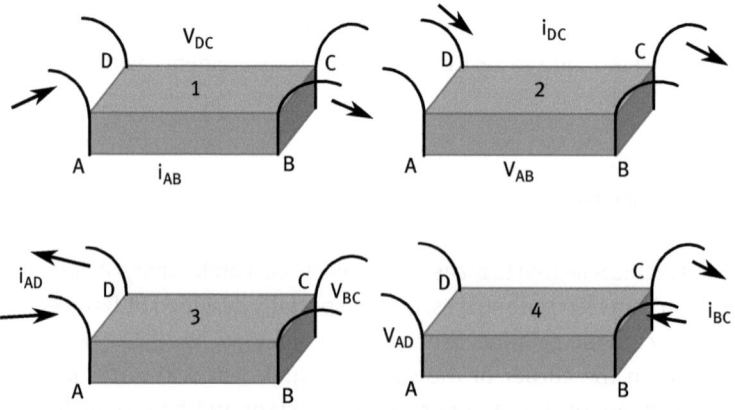

Fig. 3. Schematic illustration of four contact configurations of vdP method [2].

between electrodes D and C is made and $R_{AB,DC} = \frac{V_{DC}}{I_{AB}}$. The position of the current and voltage electrodes is then changed, so that current I_{BC} (Figs. 3 and 4) goes through electrode B and extracts through C, and the potential difference V_{AD} is between the electrodes A and D. From measured data, it is possible to get resistance $R_{BC,DA}$, which equals $\frac{V_{DA}}{I_{BC}}$ [5]. After measurements and calculations, we can find the sample resistivity from equation (1) [5]

$$\exp\left(-\pi\frac{R_{AB,CD}d}{\rho}\right) + \exp\left(-\pi\frac{R_{BC,AD}d}{\rho}\right) = 1, \tag{1}$$

where d is the thickness of the sample and ρ is the resistivity of the material.

2.4 Measurement of Gelatine Sample Conductivity

For measurements, we labeled corners of the plexiglas box A, B, C and D. If the contacts are not placed exactly on the corners of the test object, it will cause inaccuracies in the results [6]. In our research, the sample is symmetrical and in the shape of square, which reduces measurement fault. The side length (L) is 50 mm and the diameter (s) of contacts is 2 mm. The measuring error due to the size of the contacts is under 10% because s/L <0.1 [6]. For current and voltage measurements, electrodes were connected to the Wayne Kerr 6500B precision impedance analyzer [8]. The frequency range for measuring was 100 Hz to 100 MHz, and for calculations, a single frequency of 10 kHz was used.

3 Results

Several different gelatine mixture recipes were used during testing period, and many measurements were made to find conductivity of these samples.

Tab. 2 shows measurement results when different amount of NaCl was added to the gelatine samples.

Tab. 2. Measurement results

Sodium chloride (NaCl), g	$R_{AB,DC} = \frac{V_{DC}}{I_{AB}}$	$R_{BC,DA} = \frac{V_{DA}}{I_{BC}}$	$1/\rho$, S/m
0 (0 wt/vol%)	61.60Ω	43.56Ω	0.08
0.5 (0.08 wt/vol%)	27.42Ω	17.21Ω	0.19
2.5 (0.4 wt/vol%)	6.25Ω	5.37Ω	0.75
3.8 (0.6 wt/vol%)	4.04Ω	3.80Ω	0.12
5 (0.80 wt/vol%)	3.51Ω	2.36Ω	1.49

Tab. 3. Electrical conductivities of different tissues [7]

Tissue	1/ρ at 10 kHz, S/m	1/ρ at 100 kHz, S/m	1/ρ at 10 MHz, S/m
Muscle	0.34083	0.36185	0.61683
Blood	0.7000	0.7029	1.0967
Heart	0.15421	0.21511	0.50137
Body fluid	1.5	1.5	1.502

Tab. 3 gives a good comparison of typical tissue conductivities on different frequencies, and we can see that if we add 2.5 g to the sample, its conductivity is similar to blood and adding 5 g is similar to body fluid.

4 Conclusion

Results show that it is possible to develop low-cost and easily obtainable gelatine samples with similar electrical properties to biological tissues. By changing the amount of sodium chloride in the gelatine mix, we can adjust the conductivity of the phantom and verify it with measurements and vdP method.

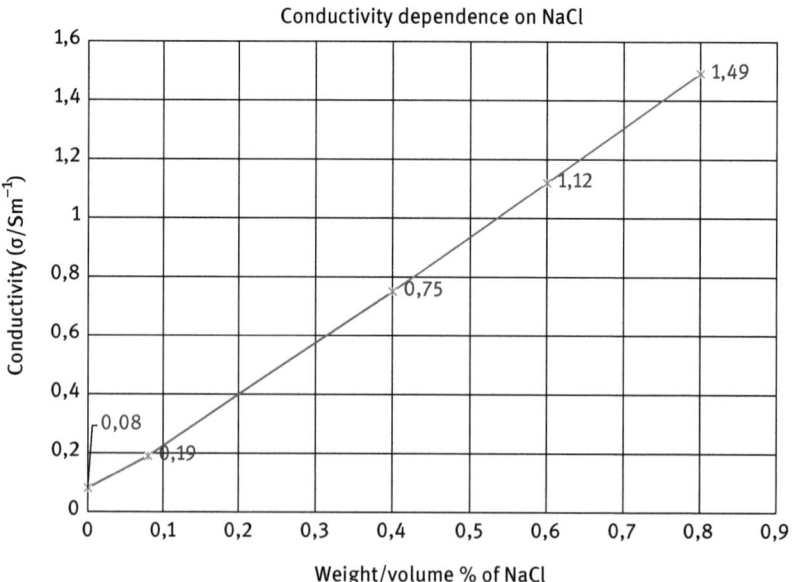

Fig. 4. Relationship between amount of sodium chloride (NaCl) added to the gelatine sample and conductivity of the sample at frequency 10 kHz.

Phantoms were measured on room temperature, and electrical conductivities of various samples were calculated on 10 kHz. After calibrating the phantoms, we have reliable and stable overtime samples to conduct research for the development of bioimpedance applications. Measurements show that conductivity of the phantom is in proportion to the amount of NaCl in the phantom composition. Also, making heterogeneous gelatine samples is currently in progress.

Acknowledgment: This research was supported by the European Union through the European Regional Development Fund in the frames of the center of excellence in research CEBE, competence center ELIKO, of the Competence Centre program of Enterprise Estonia, National Development Plan for the Implementation of Structural Funds Measure 1.1 project 1.0101.01-0480, and by Estonian Science Foundation Grant 9394.

5 References

[1] A. Hunt, A. Ristolainen, P. Ross, et al., "Low cost anatomically realistic renal biopsy phantoms for interventional radiology trainees," European Journal of Radiology, vol. 82, pp. 594–600, 2013.
[2] D. Daghero "Resistivity measurements the conventional and van der Pauw techniques," PhD thesis, Polytechnic University of Turin, 2002.
[3] T. Matsumura, Y. Sato " A theoretical study on Van der Pauw measurement values of inhomogeneous compound semiconductor thin films," Journal of Modern Physics, vol. 1, no. 5, pp. 340–347, November 2010.
[4] M. J. Burns, "Quick & dirty review of resistivity measurement techniques," MICE, final report on U.S. Government Contract No. DABT63-99-C-0016, April 2000.
[5] M. Elbohouty, M. Wilson, L. Voss, et al., "Methodology to measure the electrical conductivity of seizing and non-seizing mouse brain slices," Physics in Medicine and Biology, vol. 58, pp. 3599–613, 2013.
[6] K. D. Schroder, Semiconductor Material and Device Characterization, 3rd ed. IEEE Press, Wiley Interscience, 2006.
[7] D.Andreuccetti, R.Fossi, and C.Petrucci: An Internet resource for the calculation of the dielectric properties of body tissues in the frequency range 10 Hz - 100 GHz. Available at: http://niremf.ifac.cnr.it/tissprop/. IFAC-CNR, Florence (Italy), 1997 (Based on data published by C.Gabriel et al. in 1996).
[8] Wayne Kerr Electronics, Precision Impedance Analyzers 6500B Series. Available at: http://www.waynekerrtest.com/global/html/products/impedanceanalysis/6500.htm, 2011.
[9] D. W. Renn, "Agar and agarose: indispensable partners in biotechnology," Industrial & Engineering Chemistry Product Research and Development.
[10] C. Marchal, M. Nadi, A. J. Tosser, et al., "Dielectric properties of gelatine phantoms used for simulations of biological tissues between 10 and 50 MHz," Vol. 5, no. 6, pp. 725–732, 1989.
[11] E. L. Madsen, G. R. Frank, T. A. Krouskop, et al., "Tissue-mimicking oil-in-gelatin dispersions for use in heterogeneous elastography phantoms," Ultrasonic Imaging, vol. 25, pp. 17–38, 2003.
[12] C. Kasl and M. J. R. Hoch, "Effects of sample thickness on the van der Pauw technique for resistivity measurements," Review of Scientific Instruments, vol. 76, p. 033907, 2005.

Paco Bogónez-Franco, Pascale Pham, Claudine Gehin,
Bertrand Massot, Georges Delhomme, Eric McAdams
and Regis Guillemaud

Problems Encountered during Inappropriate Use of Commercial Bioimpedance Devices in Novel Applications

Abstract: It is often tempting to apply commercially available impedance monitoring devices to novel applications that involve impedances in the lower range for which the device was designed. This problem may be more common than expected and will result in distorted impedance loci and in the inaccurate calculation of model parameters. The authors illustrate this problem by using commercial devices, designed for whole-body bioimpedance spectroscopy measurement, in a more demanding localized impedimetric study. The lower tissue impedances involved make the measurements more prone to the adverse effects of large-contact impedances and, in particular, contact impedance mismatches. One must therefore develop novel, low-contact impedance electrodes or new devices specifically designed for more demanding localized applications.

Keywords: Commercial impedance device, electrode–skin impedance, segmental impedance, electrode mismatch

1 Introduction

A range of commercial impedance devices exist for whole-body bioimpedance analysis (BIA) in a range of clinical applications [1, 2]. Recently, there has been interest in the use of these measurements on regions or segments of the body, possibly leading to wearable or at least more convenient home use of the systems by the patients [3]. Authors are interested in the use of regional/segmental BIA measurements, non-invasive methods for detecting and evaluating changes in the hydration and nutritional status of patients with renal disease, especially for those undergoing dialysis at home.

Moving from whole body to regional/segmental BIA measurement presents some of the problems encountered in whole-body BIA measurements. Some are particular to regional/segmental BIA measurements and are reported in this chapter to warn

---

Paco Bogónez-Franco, Claudine Gehin, Bertrand Massot, Georges Delhomme and Eric McAdams, Université de Lyon, Institute des Nanotechnologies de Lyon, INL-UMR5270, Villeur- banne, Lyon, France
Pascale Pham and Regis Guillemaud, CEA-LETI MINATEC, 17 rue des Martyrs, F-38054 Grenoble, France

DOI 10.1515/9783110449822-014

others of the risks of using existing commercial devices. Many researchers used them in novel applications such as ours. In segmental BIA, it has been identified that design of the electrodes and the standardization of their placement are major concerns that limit the value of the approach. Slight differences in electrode placement can significantly change the encompassed tissue area and dramatically affect the derived parameter values. It has also been pointed out that there is a need to investigate and standardize electrode design and that this is a major problem. There is inconsistency in the type and size of electrode that minimizes contact impedance interferences and optimizes current density and distribution. Although a variety of spot electrodes are used, there is a lack of information on which type yields the best reproducibility and accuracy in prediction models. Another question is the value of spot as compared with band electrodes. Theoretically, band or circumferential electrodes should provide a homogeneous environment for optimal introduction of current and measurement of voltage drop [4].

In regional/segmental BIA measurement, there is the additional problem that the targeted tissue impedance is much smaller than in whole-body measurements, making the contact impedances due to electrodes/skin much more significant. In fact, researchers must be very careful that the impedances encountered in such regional/segmental (or other) measurement are within the operating range of the commercial device used. The authors have found that that was not the case in some of their localized impedimetric studies, even with "gold standard" laboratory devices such as the Solartron 1255B (Solartron Analytical Ltd, Hampshire, England) with 1294 interfaces and, as a result, the data obtained in these cases were grossly distorted and largely useless. This is a potential serious problem in the novel use of devices designed for whole-body measurement or for bioimpedances larger than those observed in localized measurements. Contact impedance varies between individuals and, on same person, with body site, electrode design/area, time, temperature, pressure, etc [5].

In the present study, the authors will concentrate on the problems associated with contact impedances during the use of commercial whole-body systems for the study of regional/segmental impedances. The effects of contact impedance and lead capacitance were studied by Bolton et al. [6] on three commercial devices designed for whole-body BIA. They carried out measurements on patients as well as on electrical models and found significant differences in the "tissue" impedance recorded.

Ward [7] performed body composition measurements using commercial impedance devices on volunteers and on equivalent electrical circuits. They observed that reproducibility of results obtained on volunteers was significantly lower than in electrical circuits. This result is most probably due to the variability of the electrode–skin contact impedance between subjects. The effect of electrode mismatch in the estimation of Cole parameters was investigated by Buendia et al. [8]. They found that a mismatch of impedance in the detecting electrodes led to an increment in the measurement of the impedance compared with matching electrodes.

Bogónez-Franco et al. [9] carried out a study of the effect of electrode impedance mismatch and of ground coupling on commercial impedance devices and reported that electrode impedance mismatch has a noticeable effect on measured impedance values. Electrode impedance mismatch can be caused, for example, when one of the electrodes is not well glued to the patient skin showing higher impedance than those that are well attached. One of the problems caused by electrode mismatch is the conversion of differential-mode voltages to common mode. Some researchers [10] have proposed the design of new measurement circuits in order to increase the common-mode rejection ratio of the measuring system and thus reduce the effect of common-mode voltages caused by electrode mismatch.

In the present work, we will investigate the effect of electrode–skin contact impedance on measured tissue impedances and the associated derived impedimetric parameters when two commercial whole-body BIA devices are used to make localized impedance measurements on the calf. These commercial devices are multi-frequency measurement devices that allow to perform bioimpedance spectroscopy analysis. The impedance properties of the human calf are of interest to studies in a range of potential clinical applications like monitoring hemodialysis therapy [11].

It must be noted that the devices were not designed for such an application, hence the resulting problems that will be presented.

2 Materials

2.1 Commercial Impedance Analyzers

Two commercial impedance analyzers were used in this study, BioparHom Z-Métrix (BioparHom, Bourget du Lac, France) and Impedimed SFB7 (Impedimed Ltd, Brisbane, Australia). All data were acquired using the software provided by the manufacturer. Calibration of devices was checked before each measurement using the calibration networks provided by the manufacturer.

2.2 Electrode–Skin Contact Impedance Measurement

In order to build a representative equivalent circuit model, contact impedances, due to the electrode/skin interface, were first measured on the calf of a test subject using a three-electrode configuration over the frequency range of 1 Hz to 10 kHz.

In three-electrode configuration, one electrode injects the current on the skin, a second electrode is placed in the area where the contact impedance is wanted to know and the third electrode is placed far away from the other two. This third electrode is used for the current return. Voltage is measured between the first and second electrodes.

3M 2660 (3M, Minneapolis, USA) hydro-gel electrodes, recommended by BioparHom, were used for the measurements, and no skin preparation was performed before placement of electrodes. We did not use alcohol to clean the skin because it dries more the skin and increases the contact impedance. Also we do not used abrasive creams because it abrades the skin.

All impedances at this stage were measured with Solartron 1255 and 1294A biological impedance interface. No distortion was observed because of its high-output impedance of current source.

Contact impedance measurements were carried out on five male volunteers. Tab. 1 shows the statistical data for the volunteers measured.

Tab. 1. Statistic information of five volunteers measured

Parameter	Mean	Standard deviation
Age (yr)	37.5	8.6
Weight (kg)	69.2	12.5
Height (cm)	178.2	23.4

2.3 Tissue Impedance measurement

Similarly, typical tissue impedance values for this location were measured over the range of 1 kHz to 1 MHz using the tetra-polar electrode configuration on the same five volunteers.

2.4 Electrode–Skin Contact Impedance Electrical Model

An equivalent electrical circuit model consisted of a resistor in series with the parallel combination of a capacitance and a resistor was built. The component values are shown in Fig. 1. Both resistor and capacitor values used in this circuit model were 1% and 5% in tolerance, respectively.

Re1 models the resistance of the electrode's gel, whereas Re2 and Ce2 model the impedance of the epidermis [12].

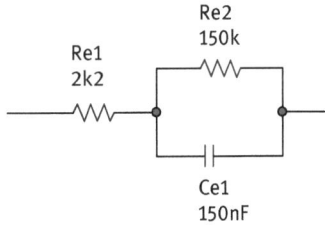

Fig. 1. Equivalent circuit model for electrode/skin contact impedance.

2.5 Tissue Electrical Model

A similar equivalent electrical circuit model that one used for electrode/skin contact impedance was built. Component values are shown in Fig. 2. Again, resistor and capacitor values used in this circuit were 1% and 5% in tolerance, respectively. This electrical model was used by Riu et al.

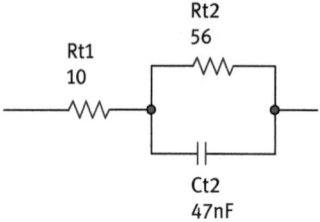

Fig. 2. Equivalent circuit model for tissue impedance.

2.6 Fitting Impedance Measurements

The frequency-dependent capacitances measured for the actual contact and tissue impedances, normally represented by the term α in empirical constant phase elements, had to be approximated by standard, frequency-independent capacitors to enable the construction of a electrical circuit model.

We used Z-View software (Scribner Associates Incorporated, North Carolina, USA) to help in this work. As the α values obtained from the fitting process were very close to 1, 0.992 for electrode–skin and 0.996 for tissue impedance, the values of the magnitude of the constant phase elements were used for the capacitance values (Ce2 and Ct2, respectively) in the electrical models.

Impedance values measured on combinations of these electrical circuits using Impedimed SFB7 and BioparHom Z-Métrix were used to assess the devices' sensitivities to such contact impedances under these novel circumstances representative of localized BIA measurements.

2.7 Errors

Absolute errors were calculated by comparing the actual equivalent circuit parameter values used for the tissue impedance and those measured by the devices in the presence of contact impedance and mismatched contact impedance circuits.

2.8 Measurements on Healthy People

In order to check the effect of electrode contact impedance mismatch, we also carried out measurements on the human calf. Electrode mismatch was produced by connecting two electrodes together at a given contact to cause a theoretical reduction in electrode/skin contact impedance of 50%. Fig. 3 shows the four-electrode configuration and the possible pairing of contact electrodes to increase the electrode impedance mismatch.

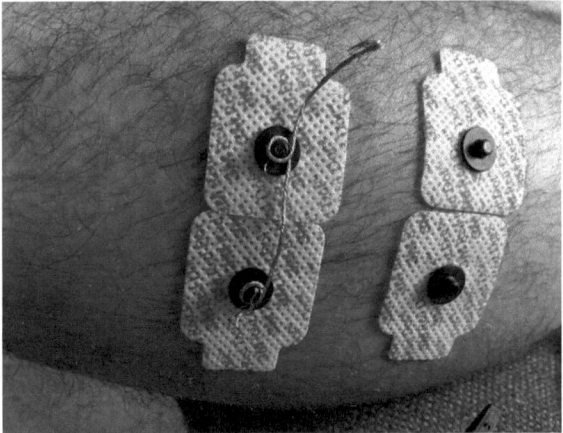

Fig. 3. Four-electrode configuration with electrode pairs used to cause electrode mismatch.

3 Experimental Results

3.1 Effect of Contact Impedance

The effect of contact impedance was evaluated using the equivalent electrical circuits described above. Over the frequencies of interest, the magnitude of contact impedance is mainly determined by resistor Re1 of Fig. 1.

In order to study the effect of contact impedance on measured tissue impedance, resistor Re1 was varied from 0 to 10 kΩ. Each lead was connected to the tissue impedance via contact impedances. Model contact and tissue impedances parameter values were those shown in Figs. 1 and 2, respectively. Tabs. 2 and 3 show the absolute errors calculated for each parameter of tissue equivalent electrical circuit model measured with BioparHom and Impedimed devices, respectively.

Tab. 2. Absolute errors in the estimation of tissue parameters for BioparHom Z-Métrix with contact impedances on injecting leads

Re1 (kΩ)	Error Rt1 (Ω)	Error Rt2 (Ω)	Error Ct2 (nF)
0	1.54	−1.02	−4.40
1	26.80	−26.17	86.00
2	*	*	*
3	*	*	*
4	*	*	*
5	*	*	*
6	*	*	*
7	*	*	*
8	*	*	*
9	*	*	*
10	*	*	*

Tab. 3. Absolute errors in the estimation of tissue parameters for Impedimed SFB7 with contact impedances on injecting leads

Re1 (kΩ)	Error Rt1 (Ω)	Error Rt2 (Ω)	Error Ct2 (nF)
0	0.15	−0.49	−2.97
1	0.62	−0.84	−1.26
2	4.50	−4.56	5.36
3	9.63	−9.69	15.39
4	14.10	−13.97	28.70
5	18.57	−18.25	41.50
6	28.98	−29.02	109.10
7	36.30	−36.36	205.60
8	44.44	−44.51	510.34
9	*	*	*
10	*	*	*

Cells filled with * in Tabs. 2 and 3 mean that the response of the tissue model measured behaves like inductive circuit, which is totally wrong in the range of frequencies measured.

3.2 Effect of Contact Impedance Mismatch

Electrode/skin contact impedance mismatch was modeled by connecting a contact circuit to one lead at time and varying the series resistance, Re1, from 1 to 10 kΩ.

3.2.1 Effect of contact impedance mismatch on positive injection leads

The positive injecting lead was connected via contact impedance to the tissue impedance circuit, whereas the remaining three leads were connected directly to the tissue circuit. Tabs. 4 and 5 show the absolute errors in the measured tissue impedance parameters as derived by the BioparHom and Impedimed devices, respectively.

Tab. 4. Absolute errors in the estimation of tissue parameters for BioparHom Z-Métrix with contact impedances on positive injecting leads

Re1 (kΩ)	Error Rt1 (Ω)	Error Rt2 (Ω)	Error Ct2 (nF)
1	0.10	0.90	−7.08
2	−0.47	1.20	−7.61
3	−2.01	2.41	−7.70
4	−3.40	3.96	−7.59
5	−6.10	6.94	−8.23
6	−7.65	8.19	−8.11
7	−8.25	9.01	−6.98
8	−10.28	10.59	−7.47
9	−10.46	10.84	−6.02
10	−10.65	10.97	−4.33

Tab. 5. Absolute errors in the estimation of tissue parameters for Impedimed SFB7 with contact impedances on positive injecting leads

Re1 (kΩ)	Error Rt1 (Ω)	Error Rt2 (Ω)	Error Ct2 (nF)
1	0.54	−0.75	−2.37
2	0.64	−0.79	−2.51
3	0.62	−0.81	−2.97
4	0.54	−0.72	−3.48
5	0.64	−0.80	−3.37
6	0.73	−0.87	−3.26
7	0.70	−0.86	−3.68
8	0.82	−0.89	−3.50
9	0.75	−0.87	−4.01
10	0.96	−1.06	−3.56

3.2.2 Effect of contact impedance mismatch on negative injection leads

The negative injecting lead was connected via contact impedance to the tissue impedance circuit, whereas the remaining three leads were connected directly to the tissue circuit. Tabs. 6 and 7 show the absolute errors in the measured tissue parameters as derived by the Impedimed and BioparHom devices, respectively.

Tab. 6. Absolute errors in the estimation of tissue parameters for BioparHom Z-Métrix with contact impedances on negative injection leads

Re1 (kΩ)	Error Rt1 (Ω)	Error Rt2 (Ω)	Error Ct2 (nF)
1	5.63	−5.08	8.47
2	9.46	−8.98	20.70
3	112.6	−12.12	28.80
4	15.00	−14.54	37.12
5	17.47	−17.11	47.98
6	19.74	−19.31	59.70
7	21.80	−21.43	73.60
8	23.80	−23.39	89.40
9	25.36	−25.05	103.90
10	26.78	−26.53	118.30

Tab. 7. Absolute errors in the estimation of tissue parameters for Impedimed SFB7 with contact impedances on negative injection leads

Re1 (kΩ)	Error Rt1 (Ω)	Error Rt2 (Ω)	Error Ct2 (nF)
1	0.21	−0.51	−2.86
2	0.30	−0.49	−2.78
3	0.28	−0.50	−2.68
4	0.21	−0.41	−2.84
5	0.19	−0.37	−2.83
6	0.14	−0.32	−2.83
7	0.09	−0.28	−2.87
8	0.06	−0.24	−2.84
9	0.00	−0.15	−3.03
10	0.01	−0.14	−2.95

3.2.3 Effect of contact impedance mismatch on positive detection leads

The positive detecting lead was connected via contact impedance to the tissue impedance circuit, whereas the remaining three leads were connected directly to the tissue circuit. Tabs. 8 and 9 show the absolute errors in the measured tissue parameters as derived by the Impedimed and BioparHom devices, respectively.

Tab. 8. Absolute errors in the estimation of tissue parameters for BioparHom Z-Métrix with contact impedances on positive detection leads

Re1 (kΩ)	Error Rt1 (Ω)	Error Rt2 (Ω)	Error Ct2 (nF)
1	−5.75	6.14	−12.55
2	−12.73	12.90	−17.75
3	−20.26	20.20	−21.35
4	−26.33	26.08	−23.62
5	−31.05	30.78	−26.61
6	−36.76	36.28	−26.08
7	−41.25	40.42	−27.58
8	−44.12	43.24	−27.45
9	−47.01	46.10	−27.18
10	−48.89	48.00	−26.57

Tab. 9. Absolute errors in the estimation of tissue parameters for Impedimed SFB7 with contact impedances on positive detection leads

Re1 (kΩ)	Error Rt1 (Ω)	Error Rt2 (Ω)	Error Ct2 (nF)
1	5.37	−5.43	7.00
2	9.95	−9.92	15.91
3	4.24	−14.52	25.84
4	7.69	−17.59	36.56
5	21.37	−21.20	51.32
6	24.89	−24.25	67.40
7	27.40	−27.31	84.40
8	30.59	−30.37	110.50
9	33.63	−33.35	147.70
10	36.32	−36.03	179.70

3.2.4 Effect of contact impedance mismatch on negative detection leads

The negative detecting lead impedance was connected via a contact impedance circuit to the tissue impedance circuit, whereas the remaining three 3 leads were connected directly to the tissue circuit. Tabs. 10 and 11 show the absolute errors in the measured tissue parameters as derived by the Impedimed and BioparHom devices, respectively.

Tab. 10. Absolute errors in the estimation of tissue parameters for BioparHom Z-Métrix with contact impedances on negative detection leads

Re1 (kΩ)	Error Rt1 (Ω)	Error Rt2 (Ω)	Error Ct2 (nF)
1	4.46	5.13	1.05
2	7.69	−7.03	7.33
3	11.08	−10.35	15.11
4	13.74	−12.90	21.34
5	16.57	−15.33	20.62
6	19.78	−18.66	39.90
7	21.08	−19.91	42.40
8	23.74	−22.50	53.70
9	25.88	−24.51	63.00
10	27.50	−26.09	70.80

Tab. 11. Absolute errors in the estimation of tissue parameters for Impedimed SFB7 with contact impedances on negative detection leads

Re1 (kΩ)	Error Rt1 (Ω)	Error Rt2 (Ω)	Error Ct2 (nF)
1	−3.65	3.59	−7.12
2	−9.84	9.60	−13.50
3	−17.20	16.67	−18.40
4	−23.22	22.81	−21.59
5	−28.39	28.09	−23.41
6	−35.32	34.91	−26.08
7	−42.48	41.94	−28.39
8	−49.64	49.00	−30.19
9	−55.99	55.40	−31.55
10	−63.61	62.93	−32.95

3.2.5 Effect of electrode impedance mismatch in in vivo measurements

Electrode contact impedance mismatched during in vivo measurements were augmented by connecting pairs of electrodes at a given lead in order to double the contact area and thus decrease the contact impedance by approximately 50%. Resultant complex impedance plots measured using the BioparHom and Impedimed devices are presented in Figs. 4 and 5. Relative errors in resistance (R) and reactance (Xc) measured were calculated for the following coincident frequencies on both devices: 5, 50,

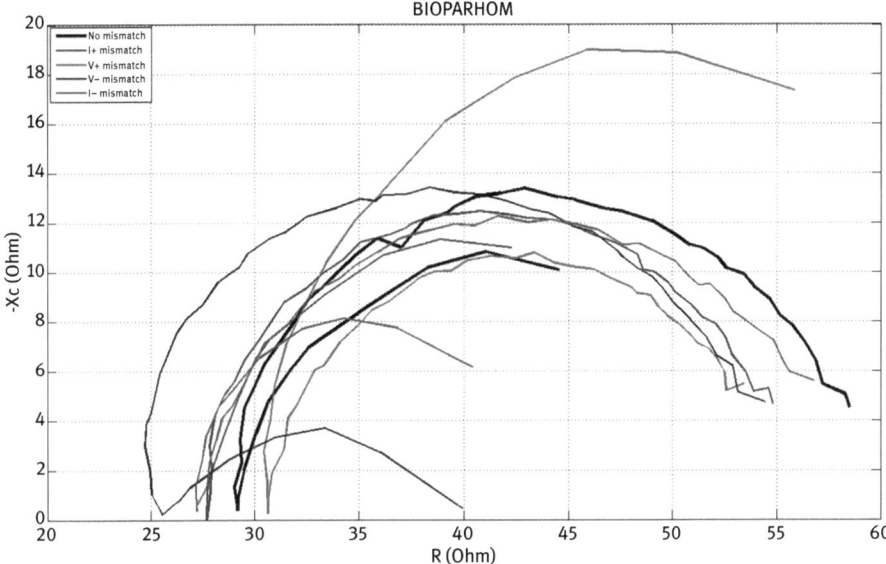

Fig. 4. Complex impedance plot for BioparHom Z-Métrix device for one of the in vivo measurements.

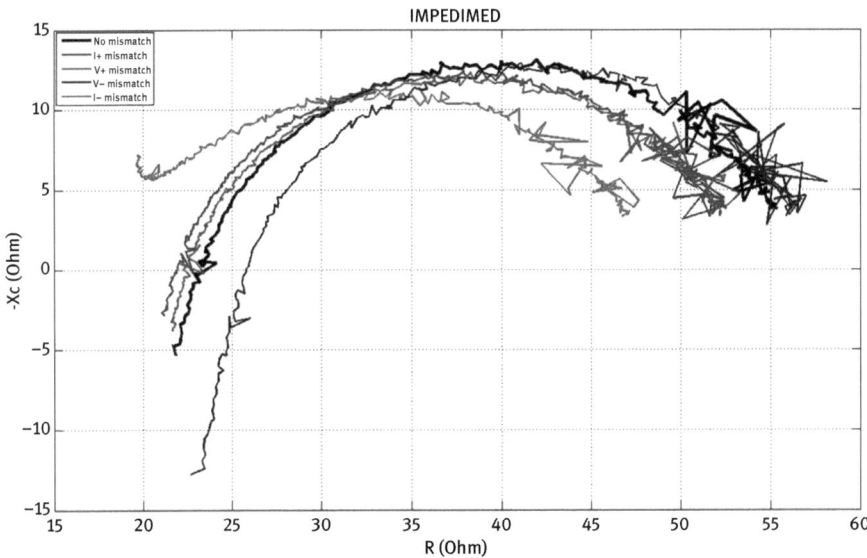

Fig. 5. Complex impedance plot for Impedimed SFB7 device for one of the in vivo measurements.

100 and 200 kHz with respect to "zero" mismatch condition (i.e., unaltered contact impedances). Tabs. 12 and 13 show the relative errors calculated for BioparHom device, respectively.

Tab. 12. Relative errors for BioparHom Z-Métrix in resistance (R) and reactance (Xc) at 5, 50, 100 and 200 kHz due to a variation in contact impedance mismatch in each lead.

lead	5	50	100	200	5	50	100	200
I+	−6.30	−8.73	−6.95	2.95	3.09	2.09	4.41	2.38
V+	−5.06	−4.99	−11.09	−5.43	−5.88	−23.30	0.31	−10.46
V−	−4.55	−2.23	−12.97	−5.84	−6.95	−38.33	6.65	−16.48
I−	−4.99	1.93	−13.94	−5.39	−6.94	−87.02	23.40	−24.49

Tab. 13. Relative errors for Impedimed SFB7 in resistance (R) and reactance (Xc) at 5, 50, 100 and 200 kHz due to a variation in contact impedance mismatch in each lead.

lead	5	50	100	200	5	50	100	200
I+	−2.88	−14.87	−2.31	−4.55	−11.46	−5.73	17.37	1.47
V+	−4.52	−15.54	−5.78	−5.12	−3.74	−14.01	−2.68	−4.89
V−	−5.54	−13.68	−5.64	−3.60	−5.28	−5.80	−3.52	−5.18
I−	−4.69	−13.90	−6.00	−3.23	0.52	21.91	−21.56	−2.95

4 Discussion

Both commercial devices are potentially adversely affected by the presence of electrode/skin contact impedances, especially under our conditions representing localized impedance measurements, for which they were not designed.

BioparHom device cannot tolerate contact impedances higher than 1 kΩ in all of its leads. Exceeding this value, the circuit measured behaves like an inductive circuit, which for biological tissue is not possible. Impedimed device can tolerate contact impedances up to 8 kΩ in each lead. As this value increases, the error increases too. Values higher than 8 kΩ have the same effect like on BioparHom; the circuit measured behaves like an inductive circuit.

Causing a mismatch on the positive injection lead causes in BioparHom a maximum error of −10.65 Ω in the estimation of Rt1, 10.97 Ω in the estimation of Rt2 and −4.33 nF in the estimation of Ct2 when the contact impedance mismatch is of 10 kΩ. Impedimed device is less affected to contact impedance mismatch on positive injection lead. Maximum errors were 0.96 Ω, −1.06 Ω and −3.56 nF in the estimation of Rt1, Rt2 and Ct2, respectively, when contact impedance mismatches.

A mismatch in the negative injection leads has bigger effect on BioparHom device than on Impedimed. The maximum errors for BioparHom device when contact impedance mismatch is 10 kΩ were 26.78 Ω, −26.50 Ω and 118.90 nF for the estimation of Rt1, Rt2 and Ct2, respectively. In the case of Impedimed, the maximum error in the estimation of Rt1 was 0.30 Ω and was caused when the contact impedance mismatch was 2 kΩ. For the case of estimation of Rt2, the maximum error was −0.51 Ω and was caused with a contact impedance mismatch of 1 kΩ. For the estimation of Ct2, the maximum error was −3.03 nF when the contact impedance mismatch was 9 kΩ. When contact impedance mismatch is caused on positive detection leads, both devices are strongly affected by their value. The maximum error for BioparHom in the estimation of Rt1 and Rt2 was −48.89 Ω and 48.00 Ω, respectively, when contact impedance mismatch was 10 kΩ. The maximum error in the estimation of Ct2 was −28.58 nF with a contact impedance mismatch of 7 kΩ. Maximum errors for Impedimed were 36.32 Ω, −36.03 Ω and 179.70 nF in the estimation of Rt1, Rt2 and Ct2, respectively, with a contact impedance mismatch of 10 kΩ.

Again, the effect of contact impedance mismatch on negative injection leads a bigger effect than on injection leads. Maximum error was achieved, on both devices, when contact impedance mismatch was 10 kΩ. For BioparHom, the maximum errors were 27.50 Ω, −26.90 Ω and 70.80 nF in the estimation of Rt1, Rt2 and Ct2, respectively. For Impedimed, these values were −63.61 Ω, 62.90 Ω and −32.95 nF, respectively.

When contact impedance mismatch was caused in in vivo measurements, we can appreciate big effect on the accuracy of the measurement if the mismatch was caused on negative injection lead and detection leads. For BioparHom, we obtained errors bigger than 10% in resistance (R) in the injection lead at 50 kHz and in negative injection and detection leads at 100 kHz. For reactance (Xc), the errors bigger than 10% were obtained on the negative injection lead at 50, 100 and 200 kHz and in the detection leads at 50 and 200 kHz.

For Impedimed, bigger errors than 10% in the measurement of the resistance were obtained in injection and detection leads at 50 kHz. For the reactance, these were obtained in positive injection lead at 5 and 100 kHz and in positive detection and negative injection leads at 50 kHz.

5 Conclusion

With localized impedimetric measurements, the tissue impedance under study is small and, as a result, more sensitive to contact impedances and contact impedance mismatch. In the present study, it was observed that the commercial, whole-body bioimpedance devices were relatively less sensitive to contact impedance but very sensitive to contact impedance mismatch.

This adverse effect was largest when the mismatch involves the detecting leads rather than the injecting leads. This effect could be reduced increasing the input impedance and common mode rejection ratio of the input stage. The influence of contact impedances must be minimized or must become negligible with respect to the magnitude of the expected changes in the monitored tissue impedance. According to the results of the present study, the magnitudes of interface impedances should be no larger than 0.5 kΩ to 1 kΩ for the segment measured.

In order to achieve this, better electrodes must be designed, possibly with larger areas, such as band electrodes, which may also help optimize the current distribution throughout the segment under investigation. The authors are presently developing novel electrode systems for their particular application.

Although it is tempting to use commercial impedance monitoring devices in novel ways to develop new and exciting applications of bioimpedimetric analysis, one must be very careful to check that the new applications do not involve impedance ranges outside those for which the devices were designed.

In the authors project, involving the development of regional/segmental BIA measurements as a method of detecting and evaluating changes in the hydration and nutritional status of home-based patients, this is unfortunately the case. The authors are also working with a manufacturer of BIA systems to develop a system specifically designed for localized impedimetric measurements, thus overcoming the problems associated with this challenging application.

Acknowledgment: This work has been supported by French National Agency Research (ANR) through TecSan (Technologies pour la santé et l'autonomie) program (Project DIALYDOM no. ANR-12-TECS-0003-01).

6 References

[1] F. Zhu, "Methods and reproducibility of measurement of resistivity in the calf using regional bioimpedance analysis," Blood Purification, vol. 21, no. 1, pp. 131–136, 2003.
[2] A. De Lorenzo and A. Andreoli, "Segmental bioelectrical impedance analysis," Current Opinion in Clinical Nutrition and Metabolic Care, vol. 6, issue 5, pp. 551–555, 2003.
[3] J. Ferreira, F. Seoane, and K. Lindercrantz, "Portable bioimpedance monitor evaluation for continuous impedance measurements. Towards wearable plethysmography applications," in 35th Annual International Conference of the IEEE, 2013, Osaka Japan, pp. 559–562.
[4] E. T. McAdams, J. Jossinet, A. Lackermeier, et al., "Factors affecting electrode-gel-skin interface impedance in electrical impedance tomography," Medical Biological Engineering, vol. 34, no. 6, pp. 397–408, 1996.
[5] M. Fernandez and R. Pallas-Areny, "Ag/AgCl electrode noise in high-resolution ECG measurements," Biomedical Instrumentation & Technology, vol. 34, no. 5, pp. 125–130, 2000.
[6] M. P. Bolton, L. C. Ward, A. Khan, et al., "Sources of error in bioimpedance spectroscopy," Physiological Measurement, vol. 19, no. 2, pp. 235–245, 1998.

[7] L. C. Ward, "Inter-instrument comparison of bioimpedance spectroscopic analyzers," The Open Medical Devices Journal, vol. 1, pp. 3–10, 2009.

[8] R. Buendia, P. Bogónez-Franco, L. Nescolarde, et al., "Influence of electrode mismatch on Cole parameter estimation from total right side electrical bioimpedance spectroscopy measurements," Medical Engineering & Physics, vol. 34, no. 7, pp. 1024–1028, 2012.

[9] P. Bogónez-Franco, L. Nescolarde, R. Bragós, et al., "Measurements errors in multifrequency bioelectrical impedance analyzers with and without impedance electrode mismatch," Physiological Measurement, vol. 30, no. 7, pp. 573–587, 2009.

[10] G. I. Petrova, "Influence of electrode impedance changes on the common-mode rejection ratio in bioimpedance measurements," Physiological Measurement, vol. 20, no. 4, pp. 11–19, 1999.

[11] U. G. Kyle, I Bosaeus, A. DeLorenzo, et al., "Bioelectrical impedance analysis–Part II: Utilization in clinical practice," Clinical Nutrition, vol. 23, no. 6, pp. 1430–1453, 2004.

Bei Fragen zur Produktsicherheit wenden Sie sich bitte an:
If you have any questions regarding product safety,
please contact:

Walter de Gruyter GmbH
Genthiner Straße 13
10785 Berlin
productsafety@degruyterbrill.com